CONGRESSIONAL POLICIES, PRACTICES AND PROCEDURES

KEY CONGRESSIONAL REPORTS FOR JUNE 2019

ENVIRONMENT

Congressional Policies, Practices and Procedures

Additional books and e-books in this series can be found on Nova's website under the Series tab.

CONGRESSIONAL POLICIES, PRACTICES AND PROCEDURES

KEY CONGRESSIONAL REPORTS FOR JUNE 2019

ENVIRONMENT

STEPHANIE A. BETTS
EDITOR

snova
New York

Library of Congress Cataloging-in-Publication Data

ISBN: 978-1-53616-662-0

Published by Nova Science Publishers, Inc. † New York

CONTENTS

PREFACE

This book is a comprehensive compilation of all reports, testimony, correspondence and other publications issued by the Congressional Research Service on Environment during the month of June.

Chapter 1 - The Land and Water Conservation Fund (LWCF) Act of 1965 was enacted to help preserve, develop, and ensure access to outdoor recreation facilities to strengthen the health of U.S. citizens. The law created the Land and Water Conservation Fund in the U.S. Treasury as a funding source to implement its outdoor recreation goals. The LWCF has been used for three general purposes. First, it has been the principal source of monies for land acquisition for outdoor recreation by four federal agencies—the Forest Service, National Park Service, Fish and Wildlife Service, and Bureau of Land Management. Second, the LWCF provides "financial assistance to states" by authorizing a matching grant program to assist states in recreational planning, acquiring recreational lands and waters, and developing outdoor recreational facilities. There are two components to this "stateside" program: the traditional formula state grants and the more recent competitive state grants. Third, beginning in FY1998, LWCF has been used to fund other federal programs with natural resource-related purposes, such as the Forest Legacy program of the Forest Service and grants under the Cooperative Endangered Species Conservation Fund of the Fish and Wildlife Service. Under the LWCF Act, the fund is authorized to accrue $900 million annually from multiple sources. However, nearly all of the revenue is derived from oil and gas leasing in the Outer Continental Shelf (OCS). The LWCF receives *additional* money under more recent legislation (P.L. 109-432, "GOMESA"). Throughout the history of the LWCF, $40.9 billion in revenues have accrued under both the LWCF Act and GOMESA. Congress determines the level of discretionary funding for the three LWCF purposes through the annual appropriations process. These discretionary appropriations have fluctuated widely since the origin of the program. In addition, any funds deposited under GOMESA are mandatory appropriations for the state grant program. Mandatory appropriations were relatively small from FY2009 to FY2017. They increased substantially beginning in FY2018 and are expected to remain relatively high over at least the next decade. Less than half of the $40.9 billion in total revenues that have accrued in the LWCF have been appropriated ($18.9 billion), leaving an unappropriated balance of approximately $22.0 billion in the fund. FY2001 marked the highest funding ever, with appropriations from the LWCF exceeding the authorized level by reaching nearly $1 billion. For FY2019, the most recent fiscal year, the total appropriation was $506.6 million (with $435.0 million in discretionary funds and $71.6 million in mandatory funds). The $18.9 billion appropriated through the history of the program has been allocated unevenly among federal land acquisition (60%), the state grant program (26%), and other purposes (14%). Similarly, federal land acquisition funds have been allocated unevenly among the four agencies. Recent Congresses have considered a variety of issues pertaining to the LWCF, through authorizing and appropriations legislation and other debates. The 116th Congress enacted several changes to the LWCF Act (in P.L. 116-9, Section 3001), including to make permanent annual deposits to the fund of $900 million and to specify portions of funding for "financial assistance to states" and land acquisitions that foster access to federal lands for recreation. Other issues remain under debate. A recurring question is the optimal level of funding for LWCF overall and its individual components. Other questions involve whether to provide additional mandatory appropriations to the LWCF; authorize LWCF funds for other programs or purposes, such as maintenance of agency lands and assets; and alter the formula for traditional stateside grants or legislatively establish criteria for competitive grants, among others.

Chapter 2 - Most of the 270 million cars, trucks, and buses on U.S. highways are powered by internal combustion engines using gasoline or diesel fuel. However, improvements in technology have led to the emergence of vehicle electrification as a potentially viable alternative to internal combustion engines. Several bills pending in the 116th Congress address issues and incentives related to electric vehicles and charging infrastructure. Experience with fully electric vehicles is relatively recent: While a few experimental vehicles were marketed in the United States in the 1990s, the first contemporary all-electric passenger vehicles were introduced in 2010. Since then, newer models have increased the range an electric vehicle can travel on a single charge, and charging stations have become more readily available. These developments have been spurred by a range of government incentives, both in the United States and abroad. Transit buses are the fastest-growing segment of vehicle electrification in China, while in the United States and the European Union, the pace of bus electrification is slower. In the United States, federal incentives for electric passenger vehicle purchases have remained largely unchanged for more than a decade and are based primarily on tax credits for electric vehicle purchases and recharging infrastructure investments, and spending on battery chemistry research to develop less-expensive technologies:

- The *plug-in electric tax credit* permits a taxpayer to take a credit of up to $7,500 for each vehicle that can be recharged from the electricity grid; it phases out after a manufacturer has sold 200,000 eligible vehicles, a threshold that has been met by Tesla and General Motors.
- A *tax credit for installation of alternative fuel vehicle refueling property* expired in 2017; it had allowed a tax credit of $1,000 for equipment installed at a residence and up to $30,000 for business installations.

- *Investment in transportation electrification research and development (R&D)*, which has led to the gradual reduction in the cost of producing lithium-ion batteries, is administered by the U.S. Department of Energy (DOE) in cooperation with private industry. Although the Trump Administration has recommended large reductions in these programs, Congress has maintained annual funding for sustainable transportation of nearly $700 million in the past two fiscal years.

Other programs that directly influence the level of vehicle electrification include the DOE Clean Cities Program, which supports local efforts to reduce fossil fuel-powered transportation, and the Department of Transportation's Alternative Fuel Corridors, which are designated Interstate Highway corridors with a sufficient number of alternative fueling stations, including electric vehicle chargers, to allow alternative fuel vehicles to travel long distances. The federal government also funds municipal transit bus electrification through Federal Transit Administration grants, which may be used for the purchase of all-electric buses. The pace of electrification also may be affected by proposals for less stringent federal standards for Corporate Average Fuel Economy (CAFE) and greenhouse gas emissions from vehicles. Beyond these federal programs, states and electric utilities provide a range of incentives for electrification. The National Conference of State Legislatures reports that 45 states and the District of Columbia offer incentives such as income tax credits for electric vehicle and charger purchases, reduced registration fees, and permitting solo drivers of electric vehicles to use carpool lanes. The California Zero Emission Vehicle program is spurring sales of electric vehicles in 10 states. Utilities can provide incentives to charge during off-peak hours, install public electric charging infrastructure, and utilize vehicle-to-grid (V2G) storage. V2G storage would allow idle vehicle batteries to supply electricity to the grid rather than drawing power from it during peak demand periods.

Chapter 3 - The Central Valley Project (CVP), a federal water project owned and operated by the U.S. Bureau of Reclamation (Reclamation), is one of the world's largest water supply projects. The CVP covers approximately 400 miles in California, from Redding to Bakersfield, and draws from two large river basins: the Sacramento and the San Joaquin. It is composed of 20 dams and reservoirs and numerous pieces of water storage and conveyance infrastructure. In an average year, the CVP delivers more than 7 million acre-feet of water to support irrigated agriculture, municipalities, and fish and wildlife needs, among other purposes. About 75% of CVP water is used for agricultural irrigation, including 7 of California's top 10 agricultural counties. The CVP is operated jointly with the State Water Project (SWP), which provides much of its water to municipal users in Southern California. CVP water is delivered to users that have contracts with Reclamation. These contractors receive varying levels of priority for water deliveries based on several factors, including hydrology, water rights, prior agreements with Reclamation, and regulatory requirements. The Sacramento and San Joaquin Rivers' confluence with the San Francisco Bay (*Bay-Delta* or *Delta*) is a hub for CVP water deliveries; many CVP contractors south of the Delta receive water that is "exported" from north of the Delta. Development of the CVP resulted in significant changes to the area's natural hydrology. However, construction of most CVP facilities predated major federal natural resources and environmental protection laws. Much of the current debate related to the CVP revolves around how to deal with changes to the hydrologic system that were not significantly mitigated for when the project was constructed. Thus, multiple ongoing efforts to protect species and restore habitat have been authorized and are incorporated into project operations. Congress has engaged in CVP issues through oversight and at times legislation, including provisions in the 2016 Water Infrastructure Improvements for the Nation (WIIN Act; P.L. 114-322) that, among other things, authorized changes to operations in an attempt to provide for delivery of more water under certain circumstances. Although some stakeholders are interested in further operational changes to enhance CVP water deliveries, others are focused on the environmental impacts of operations. Various state and federal proposals are currently under consideration and have generated controversy for their potential to affect

CVP operations and allocations. In late 2018, the State of California finalized revisions to its Bay-Delta Water Quality Control Plan. These changes would require that more flows from the San Joaquin and Sacramento Rivers reach the Bay-Delta for water quality and fish and wildlife enhancement (and thus would further restrict water supplies for other users). The changes have generally been opposed by the Trump Administration. At the same time, the Trump Administration is pursuing efforts to increase CVP water supplies for users, including changes to CVP operations under an October 2018 White House memorandum on western water supplies. Efforts to add or supplement CVP storage and conveyance are also being considered: The state is proposing a new water conveyance project (known as the California WaterFix) that would bypass the Bay-Delta and, under certain conditions, increase exports from north to south for some users. Additionally, new storage projects are under study by federal and state entities; these projects would aim to increase CVP and/or SWP water supplies. In the 116th Congress, legislators may consider bills and conduct oversight on efforts to increase CVP water exports compared to current baselines. Congress is considering whether to approve funding for new water storage projects, and may also consider legislation to extend or amend previously enacted CVP authorities (e.g., WIIN Act authorities that are expiring or have exceeded their appropriations ceiling).

Chapter 4 - The use of biomass as an energy feedstock has regularly been presented as a potentially viable alternative to address U.S. energy security concerns, foreign oil dependence, and rural economic development, and as a tool to possibly help improve the environment (e.g., through greenhouse gas emission reduction). Biomass (organic matter that can be converted into energy) includes food crops, crops grown specifically to produce energy (e.g., switchgrass or prairie perennials), crop residues, wood waste and byproducts, and animal manure. Biomass may be used to produce heat, electricity, or liquid transportation fuel. Efforts to promote the use of biomass for liquid transportation fuels have focused on corn primarily. Efforts to promote the use of biomass for power generation have focused on wood, wood residues, and milling waste primarily. Comparatively less emphasis has been placed on the use of other biomass feedstocks—noncorn food crops, nonfood crops, crop residues, animal manure, and more—as renewable energy sources for liquid fuel use or for power generation. This is partly due to the variety, lack of supply, and dispersed location of noncorn-based biomass feedstock. The technology development status and costs to convert noncorn-based biomass into energy are also viewed by some as obstacles to rapid adoption or deployment. To aid in understanding the role of biomass as an energy resource, this chapter examines the characterization of biomass in legislation. For over 40 years, the term *biomass* has been used in legislation enacted by Congress for various programs. Biomass-related legislation has provided financial incentives to develop technologies that use biomass. How biomass is defined influences decisions about the type of biomass that is grown, where it is grown, and potential preferred energy uses, among other things. There have been 14 biomass definitions included in legislation—including tax legislation—since 2004. Future policy discussions about both energy—particularly legislation involving the Renewable Fuel Standard (RFS) and energy tax incentives—and the environment may prompt further discussion about the definition of biomass. For example, one point of contention regarding the biomass definition and the RFS is whether the term should be defined to include a larger quantity of biomass from federal lands. Some argue that removal of biomass from these lands may lead to ecological harm. Others contend that biomass from federal lands can aid the production of renewable energy to meet certain mandates (e.g., the RFS) and that the removal of biomass can enhance forest restoration or protection from wildfires, insects, or diseases. Bills have been introduced—in the current Congress and previous Congresses—that would modify the biomass definition, indicating some interest in expanding or refining its use. For example, S. 1614 would expand the renewable biomass definition for the RFS to include more biomass material from federal lands. Bills that would modify the biomass definition were also introduced in the 111th, 112th, and 113th Congresses. This chapter lists biomass definitions enacted by Congress in legislation since 2004, and discusses the similarities and differences among the definitions.

Chapter 5 - President Trump and various U.S. lawmakers have expressed concerns about U.S. reliance on critical mineral imports and potential disruption of supply chains that use critical minerals for various end uses, including defense and electronics applications. Chinese export quotas on a subset of critical minerals referred to as rare earth elements (REEs) and China's 2010 curtailment of REE shipments to Japan heightened U.S. vulnerability concern. In December 2017, Presidential Executive Order 13817, "A Federal Strategy to Ensure Secure and Reliable Supplies of Critical Minerals," tasked the Department of the Interior to coordinate with other executive branch agencies to publish a list of critical minerals. The Department of the Interior published a final list of 35 critical minerals in May 2018. The concern among many in Congress has evolved from REEs and REE supply chains to include other minor minerals and metals that are used in small quantities for a variety of economically significant applications (e.g., laptops, cell phones, electric vehicles, and renewable energy technologies) and national defense applications. Also, as time passed, concerns increased about access to and the reliability of entire supply chains for rare earths and other minerals. Congressional action (e.g., National Defense Authorization Act for FY2014, P.L. 113-66) has led to the acquisition of REEs and other materials for the National Defense Stockpile. In 2017, the United States had no primary production of 22 minerals and was limited to byproduct production of 5 minerals on the critical minerals list. In contrast, the United States is a leading producer of beryllium and helium, and there is some U.S. primary production of 9 other critical minerals. China ranked as the lead global producer of 16 minerals and metals listed as critical. Although there are no single monopoly producers in China, as a nation, China is a dominant or near-monopoly producer of yttrium (99%), gallium (94%), magnesium metal (87%), tungsten (82%), bismuth (80%), and rare earth elements (80%). The United States is 100% import reliant on 14 minerals on the critical minerals list (aside from a small amount of recycling). These minerals are difficult to substitute inputs into the U.S. economy and national security applications; they include graphite, manganese, niobium, rare earths, and tantalum, among others. The United States is more than 75% import reliant on an additional 10 critical minerals: antimony, barite, bauxite, bismuth, potash, rhenium, tellurium, tin, titanium concentrate, and uranium. The current goal of U.S. mineral policy is to promote an adequate, stable, and reliable supply of materials for U.S. national security, economic well-being, and industrial production. U.S. mineral policy emphasizes developing domestic supplies of critical materials and encourages the domestic private sector to produce and process those materials. But some raw materials do not exist in economic quantities in the United States, and processing, manufacturing, and other downstream ventures in the United States may not be globally cost competitive. Congress and other decisionmakers have multiple legislative and administration options to weigh in deliberating on whether, and if so how, to address the U.S. role and vulnerabilities related to critical minerals.

Chapter 6 - The National Trails System was created in 1968 by the National Trails System Act (16 U.S.C. §§1241-1251). The system includes four types of trails: (1) national scenic trails (NSTs), which display significant physical characteristics of U.S. regions; (2) national historic trails (NHTs), which follow travel routes of national historical significance; (3) national recreation trails (NRTs), which provide outdoor recreation accessible to urban areas; and (4) connecting or side trails, which provide access to the other types of trails. As defined in the act, NSTs and NHTs are long-distance trails designated by acts of Congress. NRTs and connecting and side trails may be designated by the Secretaries of the Interior and Agriculture with the consent of the federal agency, state, or political subdivision with jurisdiction over the lands involved. Congress plays an ongoing role in shaping the National Trails System through legislation and oversight. Broad issues for Congress include, among others:

- whether and where to establish new trails in the system,
- whether to establish new trail categories (such as “national discovery trails”), and
- how much funding to provide to agencies for trail management.

When designating individual trails, Congress has considered issues such as:

- how to balance trail designation with other potential land uses,
- how to address federal land acquisition, and
- whether to make specific provisions for trail use that may differ from those authorized in the overall act.

Congress has established 11 NSTs and 19 NHTs, as well as several NRTs (although recreation trails are more typically designated administratively). In addition, the Secretaries of the Interior and Agriculture have designated just under 1,300 NRTs and seven connecting or side trails. The scenic, historic, and connecting trails are federally administered by the National Park Service (NPS) and/or the Bureau of Land Management (BLM) in the Department of the Interior, or the U.S. Forest Service (FS) in the Department of Agriculture, with cooperation from states and other entities to operate nonfederal trail segments. The roughly 1,300 national recreation trails are typically managed by states, localities, and private organizations, except where they cross federal lands. The act limits federal land acquisition for the trails system, with specific provisions for different trail types. Each federal agency with management authority over national trails has its own budget for trail administration and management. Trails have also received funding from federal transportation programs, private donations, permits and fees, and local excise taxes, among other sources. Uses of the national trails may include, but are not limited to, bicycling, cross-country skiing, day hiking, equestrian activities, jogging or similar fitness activities, overnight and long-distance backpacking, snowmobiling, and surface water and underwater activities. Provisions for motorized vehicle use vary among the different types of trails. In March 2019, the 116th Congress passed and the President signed the John Dingell, Jr. Conservation, Management, and Recreation Act (P.L. 116-9), which extended existing national scenic and historic trails and directed the study of an additional trail route for potential addition to the system. Other legislation introduced in the 116th Congress would designate a new national historic trail (H.R. 434) and establish a new tax credit for conservation contributions to NSTs (S. 809 and H.R. 1727). As in earlier Congresses, a bill (H.R. 726) has also been introduced to add a new type of trail—national discovery trails—to the system.

Chapter 7 - The share of wind and solar power in the U.S. electricity mix grew from 1% in 2008 to 8% in 2018. Wind and solar are variable renewable energy (VRE) sources. Unlike conventional sources, weather variability creates uncertainty about the availability of VRE sources. This uncertainty could potentially result in a lack of reliability. Some Members of Congress have expressed concerns about the reliability of the electric power system given recent growth in generation from wind and solar sources and projections that growth will continue. According to official metrics, electric reliability was generally stable or improving over the 2013-2017 period. In other words, generation from wind and solar sources does not appear to be causing electric reliability issues at the national level over this period. Questions remain, however, about maintaining reliability if generation from wind and solar should increase above current projections, as some Members of Congress have supported. Entities in the electric power sector and their regulators are evaluating changes to their approaches to reliability to prepare for this possibility. Congress might seek clarification on whether new or modified approaches are required. Under the current regulatory framework, the federal government oversees reliability for the generation and transmission systems of the electric power sector. These components comprise the bulk power system and include large-scale wind and solar sources. The Energy Policy Act of 2005 (EPACT05; P.L. 109-58) authorized the Federal Energy Regulatory Commission (FERC) and the North American Electric Reliability Corporation (NERC) to develop and enforce mandatory reliability standards for the bulk power system. Small-scale wind and solar sources, such as rooftop solar photovoltaic (PV) panels, are connected to the distribution system which is localized and under state jurisdiction. Federal mandatory reliability standards do not apply to the distribution system. The colloquial definition of reliability is "having power when it is needed," but regulators and operators of power system components require a more precise statement of objectives and metrics. FERC and NERC have developed numerous technical standards to address reliability. These standards apply over the range of timescales over which reliability is measured, from milliseconds to years. FERC has approved approximately 100 reliability standards to date, and new standards are developed as needed to respond to changing conditions, including increasing generation from wind and solar sources. Multiple entities spanning multiple jurisdictions work together to maintain electric reliability. For economic reasons, wind and solar sources tend to be utilized to the maximum extent possible. When their availability changes, which can happen quickly, other sources must quickly respond to maintain reliability. Typically, other sources respond by increasing or decreasing their output, an operation known as balancing. Multiple types of electricity sources are used to balance wind and solar, including some fossil fuel-fired generators, some nuclear generators, other wind and solar sources (provided sufficient transmission availability), energy storage, and demand response. Each of these has benefits and limitations. Some sources and system operations that currently support balancing have received federal financial support in the past, such as tax credits, grants to states or other entities, and Department of Energy research programs. Congress might consider continuing or expanding such support, if lawmakers believed current activities affecting reliability were insufficient. Beyond developing and enforcing reliability standards, other federal government activities affect electric reliability. For example, FERC's regulation of interstate electricity transmission can be a key determinant of how effectively different electricity sources can meet demand. FERC's regulation of the wholesale electricity markets that operate in some regions of the country may also affect reliability, because market rules can influence which individual generators are used for system balancing. Market prices directly affect project revenues, influencing the kinds of sources that are developed. Additionally, some projects and programs Congress funds support reliability by enabling technology development and providing financial support for projects that support reliability.

In: Key Congressional Reports for June 2019 ISBN: 978-1-53616-662-0
Editor: Stephanie A. Betts

Chapter 1

LAND AND WATER CONSERVATION FUND: OVERVIEW, FUNDING HISTORY, AND ISSUES (UPDATED)*

Carol Hardy Vincent

SUMMARY

The Land and Water Conservation Fund (LWCF) Act of 1965 was enacted to help preserve, develop, and ensure access to outdoor recreation facilities to strengthen the health of U.S. citizens. The law created the Land and Water Conservation Fund in the U.S. Treasury as a funding source to implement its outdoor recreation goals.

The LWCF has been used for three general purposes. First, it has been the principal source of monies for land acquisition for outdoor recreation by four federal agencies—the Forest Service, National Park Service, Fish and Wildlife Service, and Bureau of Land Management. Second, the LWCF provides "financial assistance to states" by authorizing a matching grant program to assist states in recreational planning, acquiring recreational lands and waters, and developing outdoor

* This is an edited, reformatted and augmented version of Congressional Research Service Publication No. RL33531, Updated June 19, 2019.

recreational facilities. There are two components to this "stateside" program: the traditional formula state grants and the more recent competitive state grants. Third, beginning in FY1998, LWCF has been used to fund other federal programs with natural resource-related purposes, such as the Forest Legacy program of the Forest Service and grants under the Cooperative Endangered Species Conservation Fund of the Fish and Wildlife Service.

Under the LWCF Act, the fund is authorized to accrue $900 million annually from multiple sources. However, nearly all of the revenue is derived from oil and gas leasing in the Outer Continental Shelf (OCS). The LWCF receives additional money under more recent legislation (P.L. 109-432, "GOMESA"). Throughout the history of the LWCF, $40.9 billion in revenues have accrued under both the LWCF Act and GOMESA.

Congress determines the level of discretionary funding for the three LWCF purposes through the annual appropriations process. These discretionary appropriations have fluctuated widely since the origin of the program. In addition, any funds deposited under GOMESA are mandatory appropriations for the state grant program. Mandatory appropriations were relatively small from FY2009 to FY2017. They increased substantially beginning in FY2018 and are expected to remain relatively high over at least the next decade.

Less than half of the $40.9 billion in total revenues that have accrued in the LWCF have been appropriated ($18.9 billion), leaving an unappropriated balance of approximately $22.0 billion in the fund. FY2001 marked the highest funding ever, with appropriations from the LWCF exceeding the authorized level by reaching nearly $1 billion. For FY2019, the most recent fiscal year, the total appropriation was $506.6 million (with $435.0 million in discretionary funds and $71.6 million in mandatory funds).

The $18.9 billion appropriated through the history of the program has been allocated unevenly among federal land acquisition (60%), the state grant program (26%), and other purposes (14%). Similarly, federal land acquisition funds have been allocated unevenly among the four agencies.

Recent Congresses have considered a variety of issues pertaining to the LWCF, through authorizing and appropriations legislation and other debates. The 116th Congress enacted several changes to the LWCF Act (in P.L. 116-9, Section 3001), including to make permanent annual deposits to the fund of $900 million and to specify portions of funding for "financial assistance to states" and land acquisitions that foster access to federal lands for recreation. Other issues remain under debate. A recurring question is the optimal level of funding for LWCF overall and its individual components. Other questions involve whether to provide additional mandatory appropriations to the LWCF; authorize LWCF funds for other programs or purposes, such as maintenance of agency lands and assets; and alter the formula for traditional stateside grants or legislatively establish criteria for competitive grants, among others.

INTRODUCTION

The Land and Water Conservation Fund (LWCF) Act of 1965 was enacted to help preserve, develop, and ensure access to outdoor recreation resources.[1] A main goal of the law was to facilitate participation in recreation and strengthen the "health and vitality" of U.S. citizens. The law sought to accomplish this goal by "providing funds" for federal land acquisition and for federal assistance to states for outdoor recreation, including for recreation planning, acquisition of lands and waters, and development of recreation facilities. The law created the Land and Water Conservation Fund in the Treasury as a funding source to implement its outdoor recreation goals.

Currently, the fund receives $900 million annually under the LWCF Act, but these credited monies cannot be spent unless appropriated by Congress (i.e., constituting discretionary appropriations). The level of discretionary appropriations has varied widely since the origin of the fund in 1965.

Although the LWCF Act initially did not specify the authorized level of funding, the law has been amended several times to specify and provide increasing levels of authorizations. The authorization was raised to $900 million for FY1978 and has remained at this level. In 2019, Congress made permanent provisions of the LWCF Act that provide for $900 million in revenues to be deposited in the fund annually.[2] Before this change, the authority for the fund to accrue $900 million annually had expired.[3]

---

[1] Act of September 3, 1964; P.L. 88-578, 78 Stat. 897. 54 U.S.C. §§200301 et seq. The text of the law had been codified at 16 U.S.C. §§460*l*-4 et seq. It was recodified under P.L. 113-287 to 54 U.S.C. §§200301 et seq.

[2] This change was enacted in P.L. 116-9 (§3001) on March 12, 2019, as codified at 54 U.S.C. §200302.

[3] The authority for the Land and Water Conservation Fund (LWCF) to accrue $900 million annually had expired after September 30, 2015, but was later extended to September 30, 2018, through the enactment of P.L. 114-113 (Division O, §801) on December 18, 2015.

To reach the $900 million annual total, the fund accrues revenues from three specific sources, including the federal motorboat fuel tax and surplus property sales. The fund accumulates the majority of its revenues from oil and gas leases on the Outer Continental Shelf (OCS).

The LWCF receives additional money (beyond the $900 million) under the Gulf of Mexico Energy Security Act of 2006 (GOMESA),[4] and these appropriations are mandatory.[5] Under GOMESA, the fund accrues revenue from certain OCS leasing, and these monies can be used only for grants to states for outdoor recreation. These mandatory appropriations had been relatively small through FY2017, but increased substantially in FY2018 and FY2019 and are expected to remain relatively high at least over the next decade.

In addition to making permanent the $900 million in annual revenue deposits, the 116th Congress enacted other changes to the LWCF Act. These changes included specifying the share of the funding for "financial assistance to states" and for federal acquisitions that foster access to federal lands for recreation, as discussed below. Congress continues to debate whether to make additional alterations to the operation of the fund. Alterations under discussion include whether to make all or a portion of the appropriations under the LWCF Act mandatory (rather than discretionary) and whether amend the law to authorize the fund to be used for different purposes (e.g., maintenance). These questions are being debated by authorizing committees and during consideration of the annual Interior appropriations legislation.

Perennial congressional issues include (1) deciding the total appropriation for federal land acquisition, determining the level of

---

The authority again expired after September 30, 2018, before being made permanent through the enactment of P.L. 116-9 (§3001) on March 12, 2019. Information on the expiration of this provision is included in CRS In Focus IF11198, *Land and Water Conservation Fund (LWCF): Frequently Asked Questions*, by Carol Hardy Vincent and Bill Heniff Jr. The potential impact of an expiration of this provision is discussed in CRS In Focus IF10323, *Land and Water Conservation Fund (LWCF): Frequently Asked Questions Related to Provisions Scheduled to Expire on September 30, 2018*, by Carol Hardy Vincent and Bill Heniff Jr.

4 Gulf of Mexico Energy Security Act of 2006 (GOMESA), P.L. 109-432, Division C, §105.

5 The authority for the LWCF to accrue revenues under GOMESA does not have an expiration date.

acquisition funds for each of the four agencies, and identifying which lands should be acquired; (2) deciding the level of funding for the state grant program and the extent to which the grants should be competitive; and (3) determining what, if any, purposes other than federal land acquisition and state grants should be funded through LWCF and at what level. The primary context for debating these issues traditionally has been the annual Interior appropriations legislation.

How the Fund Works

The LWCF is not a true trust fund as is generally understood in the private sector. For instance, the fund is credited with revenues totaling $900 million annually under the LWCF Act, but these credited monies cannot be spent unless appropriated by Congress, as noted. Further, interest is not accrued on the accumulated unappropriated balance that has been credited to the LWCF. While some supporters assert that the LWCF was originally intended to be a revolving fund, whereby the money would be maintained in an account separate from the General Treasury that could accrue interest, this has not been the case. Although the LWCF Act has been amended, the fund's basic purpose has not been altered. Over time, notable amendments have raised the authorization ceiling, made the ceiling permanent at $900 million, and mandated that offshore oil and gas leasing revenues should make up any shortfall from other specified financing sources, among other changes.

From FY1965 through FY2019, about $40.9 billion has been credited to the LWCF under both the LWCF Act and GOMESA. To date, nearly all of the monies have accrued to the fund under the LWCF Act. Specifically, of the $40.9 billion that has accrued from FY1965 to FY2019, an estimated $264.4 million was derived from receipts under GOMESA.

Of the total credited to the LWCF, less than half—$18.9 billion—has been appropriated, leaving an unappropriated balance of approximately

$22.0 billion in the fund.[6] To date, nearly all of the appropriations have been discretionary funds provided under the LWCF Act. Specifically, of the $18.9 billion that has been appropriated from FY1965 to FY2019, an estimated $146.6 million was mandatory funding under GOMESA.[7] (See Figure 1.[8])

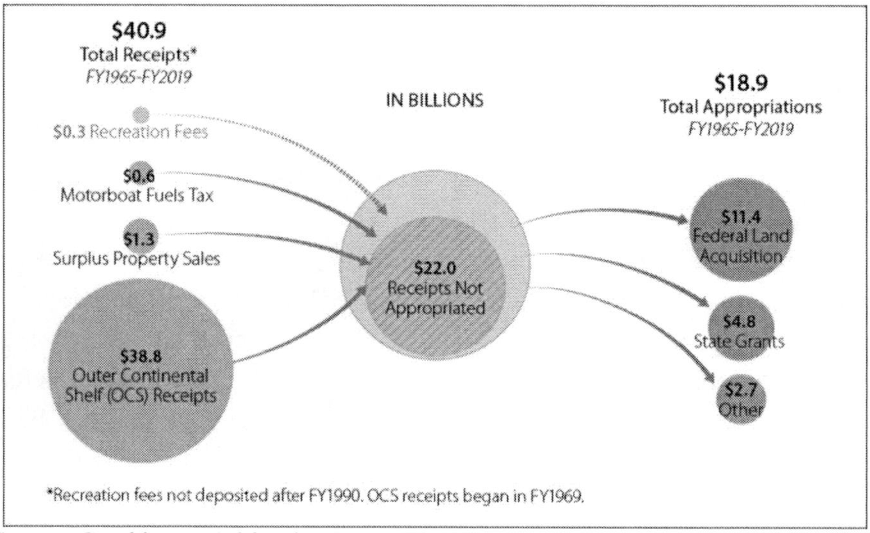

Source: Graphic created by the Congressional Research Service (CRS). The primary source for the data is the Department of the Interior (DOI) Office of Budget. Additional sources of information include the annual DOI Budget in Brief and congressional documents accompanying the annual Interior appropriations bill.

Notes: OCS receipts and appropriations for state grants reflect monies derived under the Gulf of Mexico Energy Security Act of 2006 (GOMESA; P.L. 109-432, Division C, §105). Recreation fees were not deposited into the fund after FY1990 and are thus shown differently from other revenues that currently accrue to the fund.

Figure 1. LWCF Total Receipts and Appropriations, FY1965-FY2019. (in billions of dollars, not adjusted for inflation)

---

[6] Figures on total receipts, total appropriations, and the unappropriated balance were derived primarily from data provided by the DOI Office of Budget. Additional sources of information include the annual DOI *Budget in Brief* and congressional documents accompanying the annual Interior appropriations bill.

[7] Under GOMESA, receipts credited to the LWCF in one year are appropriated from the LWCF in the following year. For this reason, the receipts through FY2019 (of $264.4 million) are higher than the appropriations through FY2019 (of $146.6 million).

[8] In this figure and throughout this chapter, amounts may not sum to totals shown due to rounding.

PURPOSES OF LWCF APPROPRIATIONS

Appropriations from LWCF have been made for three general purposes: (1) federal acquisition of land and waters and interests therein,[9] (2) the state grants for recreational planning; acquiring recreational lands, waters, or related interests; and developing outdoor recreational facilities; and (3) related purposes.[10] Each year, Congress determines the total appropriations from the fund, and the amount provided for each of these three general categories.

Division Between Federal and State Purposes

As amended in 2019, the LWCF Act states that of the total made available through appropriations or deposits under GOMESA, not less than 40% is to be used for "federal purposes" and not less than 40% is to be used to provide "financial assistance to states."[11] The 2019 amendments reinstated an explicit provision of funding for states and, for the first time, identified the total funding as including GOMESA revenues. The "federal purposes" in the LWCF Act essentially relate to land acquisition.[12] The "financial assistance to states" program set out in the LWCF Act provides for state grants for outdoor recreation (as referenced above).[13] As noted, appropriations from the LWCF have been provided for additional programs and activities, including other federal programs and types of grants to states (as discussed in the "Other Purposes" section below).

Provisions of the LWCF Act providing for a distribution of funding between federal and state activities have evolved over time. As enacted in 1964, the law generally had provided for the appropriations from the

---

[9] An interest in land is something less than full ownership, such as an easement or mineral, timber, or water rights.

[10] Hereinafter, these categories of funding often are referred to respectively as (1) federal land acquisition, (2) the stateside program, and (3) other purposes.

[11] 54 U.S.C. §200304. This text was enacted in P.L. 116-9 (§3001) on March 12, 2019.

[12] 54 U.S.C. §200306.

[13] 54 U.S.C. §200305.

LWCF to be 60% for state purposes and 40% for federal purposes. The 1964 law stated that "in the absence of a provision to the contrary in the Act making an appropriation from the fund," the appropriation from the fund was to be 60% for state purposes and 40% for federal purposes. It further specified that during the first five years in which appropriations were made from the fund, the President could vary these percentages by not more than 15 percentage points to meet the needs of states and the federal government.

In 1976, the LWCF Act was amended to state that not less than 40% of the appropriations from the fund were to be available for federal purposes. The amendment removed the text specifying 60% of the appropriations for state purposes.[14] In the conference report on the bill (S. 327), conferees expressed that "Generally, appropriations should continue to reflect the 60-40 allocation established by the Act." However, they noted the "inflexibility" of this division because "States may sometimes be unable to provide the amounts necessary to match their share of the appropriations from the fund," and thus might not be able to use their funding in a given year. In that case, additional funding should be provided to the federal agencies for acquisition in order "to preserve and protect" areas for future generations, according to the conferees.[15] At the time of the 1976 amendment, funds were being appropriated for federal land acquisition and the stateside program. Funding for other purposes did not occur until FY1998. Following the 1976 changes, no subsequent amendments were enacted on the federal/state funding distribution until 2019.

Federal Land Acquisition

The LWCF is the principal source of funds for federal acquisition of lands. Most federal lands are acquired (and managed) by four agencies—the Forest Service (FS) in the Department of Agriculture, and the National Park Service (NPS), Fish and Wildlife Service (FWS), and Bureau of Land

---

[14] These 1976 changes were contained in P.L. 94-422.

[15] See H.Rept. 94-1468 on S. 327, p. 16.

Management (BLM) in the Department of the Interior (DOI). These four agencies manage about 95% of all federally owned lands.[16] Of these agencies, the FWS has another significant source of acquisition funding. Specifically, under the Migratory Bird Conservation Fund the FWS has a permanently appropriated source of funding for land acquisition.[17] The BLM also has authority to keep the proceeds of certain land sales and use them for subsequent acquisitions and other purposes.

The LWCF Act provides that "unless otherwise allotted in the appropriation Act making them available," appropriations from the fund for federal purposes are to be allotted by the President for certain activities.[18] These activities include land acquisition in recreation areas administered by the Secretary of the Interior for recreational purposes; land acquisition in national park, national forest, and national wildlife refuge system units; and land acquisitions that foster access to federal land for recreational purposes.[19] In practice, the appropriations acts typically specify the federal purposes for which the funds are to be used.

In many respects, the process for appropriating funds for federal land acquisition is similar from year to year. The Administration's annual budget submission for each of the four major federal land management agencies generally identifies lands (or interests in lands) each agency seeks to acquire with LWCF funds. Each agency chooses among many potential options in making its annual request, through a three-level process involving identification and prioritization of acquisition projects by field, regional, and headquarters offices. In the past, the agencies used their own criteria and processes to identify and prioritize acquisition project requests. In 2019, Congress enacted common criteria for the agencies to take into

---

[16] For an overview of these agencies, see CRS In Focus IF10585, *The Federal Land Management Agencies*, by Katie Hoover. For information on the extent of federal land ownership by these agencies and the Department of Defense, see CRS Report R42346, *Federal Land Ownership: Overview and Data*, by Carol Hardy Vincent, Laura A. Hanson, and Carla N. Argueta.

[17] For more information on the Migratory Bird Conservation Fund, see the FWS land acquisition section of CRS Report RL34273, *Federal Land Ownership: Acquisition and Disposal Authorities*, by Carol Hardy Vincent et al.

[18] 54 U.S.C. §200306(a).

[19] The LWCF Act also identifies "capital costs" for recreation and fish and wildlife at water resources development projects as a federal purpose.

account in considering which lands to acquire. These criteria include the significance and urgency of the acquisition, management efficiencies and cost savings, geographic distribution, land threats, and recreational value.[20]

In the past, the Administration sometimes sought a portion of funding for certain types of acquisitions, such as those that would facilitate access to federal lands for recreation and sportsmen.[21] In 2019, Congress amended the LWCF Act to specify that a certain amount or percentage of LWCF appropriations are to be used each year for acquisitions that facilitate recreational access to federal lands. Relatedly, Congress directed the agencies to develop priority lists of acquisitions to facilitate access for hunting, fishing, recreational shooting, or other outdoor recreation purposes.[22] In some years, the Administration has not sought funding for individual land acquisition projects in the annual budget submissions to Congress, as was the case in the Trump Administration's annual budget submissions for each of FY2018-FY2020. Nevertheless, for each of FY2018- FY2020, Congress directed the Trump Administration to submit prioritized acquisition project lists for consideration during the appropriations process, and project lists were subsequently submitted to Congress. For instance, in the joint explanatory statement accompanying the FY2019 Interior appropriations law, conferees directed agencies to "continue their longstanding process of identifying and prioritizing potential Federal land acquisition projects in anticipation of program appropriations as consistent with previous years." Conferees further directed the agencies to submit prioritized acquisition project lists for FY2020, by the earlier of March 1, 2019, or 30 days after their budget submissions.[23] In accordance with this direction, the Administration sent

---

[20] P.L. 116-9 (§3001) enacted on March 12, 2019, and codified at 54 U.S.C. §200306(d).

[21] For example, such funding was sought by the Obama Administration and appropriated for FY2017 (among other years).

[22] Specifically, the law provided for not less than $15.0 million or 3% of amounts made available for expenditure each year (under 54 U.S.C. §200303), whichever is greater, to be used for recreational public access. See P.L. 116-9 (§3001) enacted on March 12, 2019, and codified at 54 U.S.C. §200306(c).

[23] H.Rept. 116-9, Conference Report to Accompany H.J.Res. 31, Making Further Continuing Appropriations for the Department of Homeland Security for Fiscal Year 2019 and for Other Purposes, p. 708.

project lists for each of the four agencies, ranging from 8 acquisitions for BLM to 25 acquisitions for the FS.

Congress reviews agency acquisition requests, determines the total acquisition funding for each agency, and decides on the portion of funding (if any) for requested types of acquisition (e.g., access for recreation and sportsmen). Congress also typically identifies (in appropriations report language) the level of funding for each acquisition project sought by the Administration.

The LWCF Act restricts appropriations to those acquisitions that have been previously authorized by law. However, it allows LWCF appropriations to be used for pre-acquisition work where "authorization is imminent and where substantial monetary savings could be realized."[24]

Appropriations laws typically provide that LWCF funds for land acquisition remain available until expended, meaning the funds can be carried over from fiscal year to fiscal year. Often an appropriation is not used in the fiscal year provided, because the process for completing a land acquisition has many components and often takes more than one year. These components generally include an appraisal of the value of the parcels, evaluations of the resources contained within the parcels, and title research.

The Administration typically seeks, and appropriations laws have typically provided, a portion of each agency's acquisition funding for one or more related activities. For instance, funds have been provided for acquisition management, costs of land exchanges, and acquisition of lands within the boundaries of federal land units ("inholdings") that may become available throughout the year. Further, in some cases funds have been appropriated for "emergencies" or "hardships," for acquisition of lands from an owner who must sell quickly and where the agency determines there is a need to purchase the lands quickly.[25]

---

[24] 54 U.S.C. §200306(b).

[25] In addition, in FY2019 and in other recent years, a portion of the NPS appropriation has been specified for the American Battlefield Protection Program for grants for nonfederal acquisition of lands (and interests) in eligible battlefields. For additional information on these acquisition grants, see the NPS website at http://www.nps.gov/abpp/grants/grants.htm. Also in FY2019 and in other recent years, a portion of the FWS

Stateside Program

Another portion of the LWCF consists of grants to states (including the District of Columbia and U.S. territories) for outdoor recreation purposes. Grants are provided for outdoor recreation purposes only rather than for indoor facilities, such as community centers. Under law, federal payments to states generally are limited to 50% or less of a project's total costs. The state project sponsor bears the remaining cost.

There are two types of grants: traditional (formula) and competitive. Both programs are administered by the NPS. Under the traditional grant program, both discretionary and mandatory appropriations are divided among states by the Secretary of the Interior (acting through the NPS). Under the competitive grant program, begun in FY2014, the NPS awards discretionary grants to urbanized areas meeting certain criteria. Under both grant programs, states may "sub-award" grants to state agencies, local units of government, and federally recognized Indian tribes.

Traditional Formula State Grants

The traditional grant program provides matching grants to states for outdoor recreation planning, acquisition of lands and waters, and facility development. More than 42,000 grants have been provided throughout the history of the program (since FY1965), with a majority used for developing new recreation facilities or redeveloping older facilities. Acquisitions funded through LWCF state grants must remain in recreation use in perpetuity, unless the Secretary of the Interior approves of the conversion of the land to another use and acceptable replacement lands are substituted. Conversions occur due to changing state or local needs, such as to use park lands to build schools, widen roads, and develop civic facilities. The NPS approves roughly 50-75 conversions yearly nationwide, typically involving a portion of the area funded with an LWCF state grant.

---

appropriation has been specified for Highlands Conservation Act grants, which provide federal matching funds for land acquisition by a state entity for land protection in partnership with the federal government. For additional information on these grants, see the U.S. government grants website at https://www.grants.gov/web/grants/search-grants.html?keywords=highlands.

The Secretary of the Interior apportions the appropriation for state grants in accordance with a formula set out in the LWCF Act.[26] The formula calls for a portion of the appropriation to be divided equally among the states.[27] The remaining appropriation is to be apportioned based on need, as determined by the Secretary.[28] Under law, the determination of need is to include the population of the state relative to the population of the United States, the use of outdoor recreation resources within a state by people outside the state, and the federal resources and programs within states. In current practice, population is the biggest factor in determining state need. No state can receive more than 10% of the total appropriation.

Discretionary appropriations for traditional state grants typically are included in the annual Interior appropriations law, without earmarks or other directions to the NPS to guide how these funds should be distributed or spent. Appropriations laws provide for these funds to be available until expended. However, under the LWCF Act, states have up to three years to use the appropriations—the federal fiscal year in which the apportionment is made and the next two fiscal years. It is rare for a state not to use the money during this time, according to the NPS. Under law, the Secretary is to reapportion among states any amount that is not paid or obligated during the three-year period.[29]

In addition, mandatory appropriations are provided for the traditional state grants only. Specifically, under GOMESA, 12.5% of the revenues

---

[26] 54 U.S.C. §200305(b).

[27] Specifically, the law provides that 40% of the first $225.0 million, 30% of the next $275.0 million, and 20% of all additional appropriations are to be apportioned equally among the states. Previously, the LWCF Act had provided that the District of Columbia and the territories be considered one state under this allocation. This provision was removed in 2019 by the enactment of P.L. 116-9 (§3001) on March 12, 2019.

[28] The apportionment among states (including the District of Columbia and U.S. territories), for FY2002 through FY2016, is on the NPS website at https://www.nps.gov/subjects/lwcf/statefundingstatus.htm. The apportionment among states for FY2017 is on the DOI website at https://www.doi.gov/pressreleases/secretary-zinke-announces-943-million-states-parks-and-outdoor-recreation-through-land. The apportionment of mandatory appropriations for FY2018 is on the NPS website at https://www.nps.gov/orgs/1207/secretary-zinke-announces-over-61-million-for-states-tosupport-parks-and-outdoor-recreation.htm. The apportionment of discretionary appropriations for FY2018 is on the DOI website at https://www.doi.gov/pressreleases/secretary-zinke-announces-100-million-support-state-parks-andoutdoor-recreation.

[29] 54 U.S.C. §200305(b)(4)).

from certain OCS leasing in the Gulf of Mexico is directed to the stateside program to be used in accordance with the terms of the LWCF Act. States generally can receive up to $125.0 million in annual stateside appropriations through FY2056, except that the maximum is $162.5 million for each of FY2021 and FY2022. The money is to be in addition to any discretionary amounts appropriated by Congress for the LWCF.

These funds also are available until expended but are subject to reapportionment among states (under the LWCF Act) if not paid or obligated after three years.

To be eligible for a grant, a state must prepare and update a statewide outdoor recreation plan. This plan must address the needs and opportunities for recreation and include a program for reaching recreational goals. It generally does not include specific projects. Under law, the plan is required to be approved by the Secretary; this responsibility has been delegated to the NPS. The states award their grant money through a competitive, open project selection process based on their recreation plans and their own priorities and selection criteria. They can use the money for state projects or for pass-through to localities or tribes. States send their ranked projects to the NPS for formal approval and obligation of grant money.[30]

Competitive State Grants

In FY2014, Congress first appropriated a portion of the LWCF (discretionary) stateside funds for a new nationally competitive outdoor recreation grant program to be developed by the NPS. Congress has continued to fund a competitive program each year, with funding increasing from $3.0 million in FY2014 to $20.0 million in FY2019. Currently, the NPS's "Outdoor Recreation Legacy Partnership Program" provides grants to states for land acquisition and development for outdoor recreation projects that are in or serve densely settled areas with

---

[30] For more information on the stateside program, see the Land and Water Conservation Fund State Assistance Program: Federal Financial Assistance Manual on the NPS website at https://www.nps.gov/subjects/lwcf/upload/lwcf_manual.pdfhttps://www.nps.gov/subjects/lwcf/upload/lwcf_manual.pdf.

populations of 50,000 or more.[31] Priority is given to communities that are underserved in terms of outdoor recreation opportunities and that have sizable populations who are economically disadvantaged. Under the 2018 grant announcement, other factors for project selection include the extent to which the projects directly connect people to the outdoors; create jobs; engage community members; involve public-private partnerships; and require coordination among the public, government, and private sector.[32] The NPS expected to issue between 25 and 35 grants (in total) with a combined sum of $13.3 million,[33] with individual grants ranging from $250,000 to $750,000. In addition to meeting the requirements in the grant announcement, projects are required to comply with the LWCF Act and other program requirements that apply to the traditional state grants. Such requirements include a nonfederal funding match (generally of at least 50% of project costs), and land use for outdoor recreation in perpetuity except with the approval of the Secretary of the Interior, as discussed above.

Other Purposes

As noted above, the LWCF Act lists the federal purposes to which the President is to allot LWCF funds "unless otherwise allotted in the appropriation Act making them available."[34] A portion of the LWCF appropriation has been provided for other federal purposes (i.e., other than land acquisition) in FY1998 and each year since FY2000.

---

[31] Urbanized areas consisting of 50,000 people or more are identified by the Bureau of the Census as part of the decennial census. The most recent list, following the 2010 census, was published in the *Federal Register* on March 27, 2012, at https://www.federalregister.gov/documents/2012/03/27/2012-6903/qualifying-urban-areas-for-the-2010-census.

[32] An April 13, 2018, grant announcement is on the U.S. government grants website at https://www.grants.gov/web/grants/search-grants.html?keywords=outdoor%20recreation%20legacy.

[33] This total reflects an appropriation of $12.0 million for FY2017 and some unused funds from FY2016.

[34] 54 U.S.C. §200306(a).

Because there is no set of "other purposes" specified to be funded from LWCF, Presidents have sought funds for a variety of purposes and Congress has chosen which, if any, other purposes to fund from LWCF. For instance, for FY2008, President George W. Bush sought LWCF funds for 11 programs within the FWS, FS, and other agencies, and Congress provided funding for two of these programs. Since FY1998, the LWCF has been used for an array of other purposes related to lands and resources, such as maintenance of agency facilities (including deferred maintenance), the Historic Preservation Fund, the Payments in Lieu of Taxes program, the FS Forest Legacy program, FWS State and Tribal Wildlife grants, and FWS Cooperative Endangered Species grants. Since FY2008, however, funds have been appropriated annually for grants under two "other" programs: Forest Legacy, and the Cooperative Endangered Species Conservation Fund.[35]

FUNDING HISTORY

Overview of Total (Discretionary and Mandatory) Appropriations, FY1965-FY2019

Nearly all of the $18.9 billion appropriated throughout the history of the LWCF was discretionary funding. Specifically, $146.6 million (0.8%) of the total was mandatory funding under GOMESA, as noted above.

The $18.9 billion appropriated through FY2019 has been unevenly allocated among federal land acquisition, the stateside program, and other purposes, as shown in Figure 2. The largest portion of the total—$11.4 billion (60%)—has been appropriated for federal land acquisition.

---

[35] Although these two programs also received LWCF funding before FY2008, they are the only two "other" programs that have received LWCF appropriations since that year. For information on "other purposes" funded by LWCF throughout its history, see CRS Report R44121, *Land and Water Conservation Fund: Appropriations for "Other Purposes,"* by Carol Hardy Vincent.

The four federal land management agencies have received differing portions of this $11.4 billion. Specifically, the NPS has received $4.7 billion (41%); the FS, $3.1 billion (27%); the FWS, $2.5 billion (22%); and the BLM, $1.0 billion (9%).[36]

The stateside program has received the second-largest portion of LWCF appropriations—$4.8 billion (26% of the total). In the early years, more funds generally went to the stateside program than to land acquisition of the four federal agencies combined. For instance, stateside appropriations exceeded federal land acquisition appropriations during 12 of the 16 years from FY1965 to FY1980. After the mid-1980s, the stateside program declined as a portion of total LWCF appropriations, and received no appropriations (except for program administration) from FY1996 through FY1999. Over the last 20 years (FY2000-FY2019), the program received less than 20% of each year's total LWCF appropriations, except in FY2002 and FY2005 and during the most recent four years (FY2016-FY2019). Of these years, the largest percentage of funds for the stateside program was 39% for FY2019, in part due to the relatively large mandatory appropriation in that year.[37] Over the last 10 years (FY2010-FY2019), stateside funding has averaged 23% of total appropriations for the LWCF.[38]

Other purposes have received the remaining portion of total LWCF appropriations—$2.7 billion (14%). No funds were provided for other purposes until FY1998. By contrast, 27% of total LWCF appropriations from FY1998 through FY2019 have been for other purposes. The FWS and FS have received the largest shares: about $1.4 billion and $1.0 billion, respectively, of the $2.7 billion appropriated for other purposes since FY1998.

---

[36] Another $0.1 billion (1%) of the total appropriations for federal land acquisition was provided to other agencies or offices (e.g., to the Office of the DOI Secretary, for land appraisal services).

[37] The stateside program received $124.0 million (29%) of the LWCF discretionary appropriations for FY2019 ($435.0 million), but $195.6 million (39%) of total LWCF appropriations (discretionary and mandatory) for FY2019 ($506.6 million).

[38] Discretionary appropriations for the stateside program were 20% of LWCF discretionary appropriations from FY2010 to FY2019. Discretionary and mandatory appropriations for the stateside program were 23% of total LWCF appropriations (discretionary and mandatory) during this 10-year period.

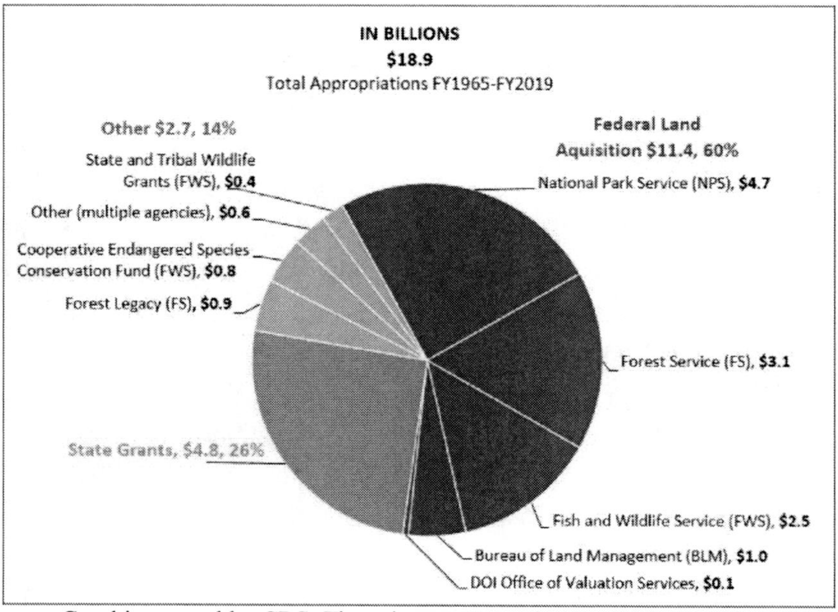

Source: Graphic created by CRS. The primary source for the data is the DOI Office of Budget. Additional sources of information include the annual DOI Budget in Brief and congressional documents accompanying the annual Interior appropriations bill.

Note: State grants reflect monies appropriated under the Gulf of Mexico Energy Security Act of 2006 (GOMESA; P.L. 109-432, Division C, §105).

Figure 2. LWCF Total Appropriations by Type and Agency, FY1965-FY2019. (In billions of dollars, not adjusted for inflation)

Overview of Mandatory Appropriations, FY2009-FY2019

Mandatory appropriations totaling $146.6 million were provided to the LWCF stateside program during an 11-year period—FY2009-FY2019. Of this total, $8.4 million was first appropriated to the stateside program in FY2009, from proceeds from OCS leasing collected in FY2008.[39]

[39] OCS receipts collected in one fiscal year are appropriated from the LWCF in the following fiscal year.

Thereafter, disbursements to the stateside program decreased, exceeding $1 million in only one other year through FY2017. The appropriation for FY2018 increased significantly—to $62.6 million—as a result of additional revenue from expansion of the qualified program areas in the Gulf of Mexico. (See *Table 1* and *Figure 3*.) The estimated appropriation for FY2019—$71.6 million—also was high relative to the prior years, and DOI anticipates that mandatory appropriations will remain high in the coming years. For instance, the NPS estimated appropriations to the stateside program at between $113 million and $125 million for each year from FY2020 through FY2025.[40] However, these estimates are subject to change, based on actual collections.

Table 1. LWCF Annual Mandatory Appropriations, FY2009-FY2019 (in millions of dollars, not adjusted for inflation)

Fiscal Year	Mandatory Appropriations
2009	$8.4
2010	0.9
2011	0.3
2012	0.1
2013	0.1
2014	1.4
2015	0.8
2016	0.1
2017	0.3
2018	62.6
2019	71.6

Source: Annual budget justifications of the NPS at https://www.nps.gov/aboutus/ budget.htm.

Notes: Under GOMESA, OCS receipts collected in one fiscal year are appropriated from the LWCF in the following fiscal year. This table shows the year in which the monies were appropriated. Figures are actual except that the FY2019 figure is estimated. Dollars represent the appropriations in the fiscal years indicated and thus are not adjusted for inflation.

[40] U.S. Department of the Interior, *Office of the Secretary and Department-Wide Programs, Budget Justifications and Performance Information, Fiscal Year 2020*, p. MLR-15, at https://www.doi.gov/sites/doi.gov/files/uploads/ fy2020_os_budget_justification.pdf.

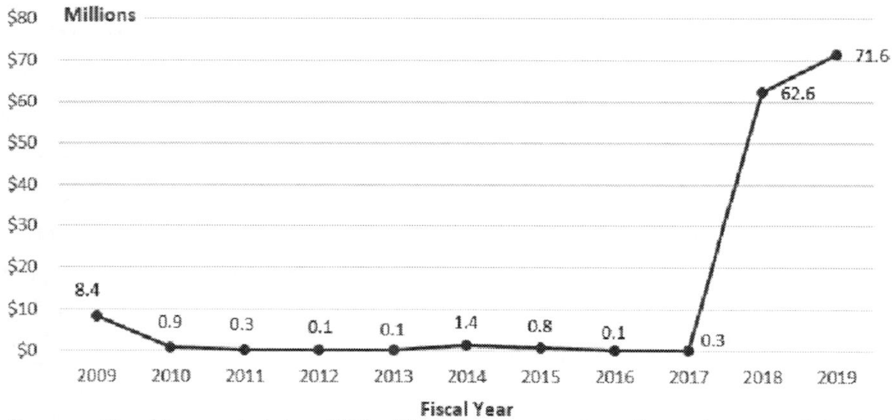

Source: Graphic created by CRS. This data was taken from the annual budget justification of the NPS at https://www.nps.gov/aboutus/budget.htm.

Notes: Under GOMESA, OCS receipts collected in one fiscal year are appropriated from the LWCF in the following fiscal year. This graphic shows the year in which the monies were appropriated. Figures are actual except that the FY2019 figure is estimated. Dollars represent the appropriations in the fiscal years indicated and thus are not adjusted for inflation.

Figure 3. LWCF Annual Mandatory Appropriations, FY2009-FY2019. (in millions of dollars, not adjusted for inflation)

Overview of Discretionary Appropriations, FY1965-FY2019

Annual discretionary appropriations from the LWCF have fluctuated widely since the origin of the program more than 50 years ago. (See Figure 4 and Table 2.) From the origin of the LWCF in FY1965 until FY1998, the discretionary total seldom reached $400 million, surpassing this level only in four years. Specifically, from FY1977 to FY1980, funding varied between $509 million (in FY1980) and $805 million (in FY1978), and averaged $647 million annually. By contrast, annual discretionary appropriations reached or exceeded $400 million in 12 of the 22 most recent years—FY1998 through FY2019.

In FY1998, LWCF discretionary appropriations spiked—to $969 million—from the FY1997 level of $159 million. FY1998 was the first year that LWCF appropriations exceeded the level the fund was authorized to accumulate in a given year ($900 million).[41] The total included $270 million in the usual funding titles for land acquisition by the four federal land management agencies; an additional $627 million in a separate title, funding both the acquisition of the Headwaters Forest in California and New World Mine outside Yellowstone National Park; and $72 million for other purposes.

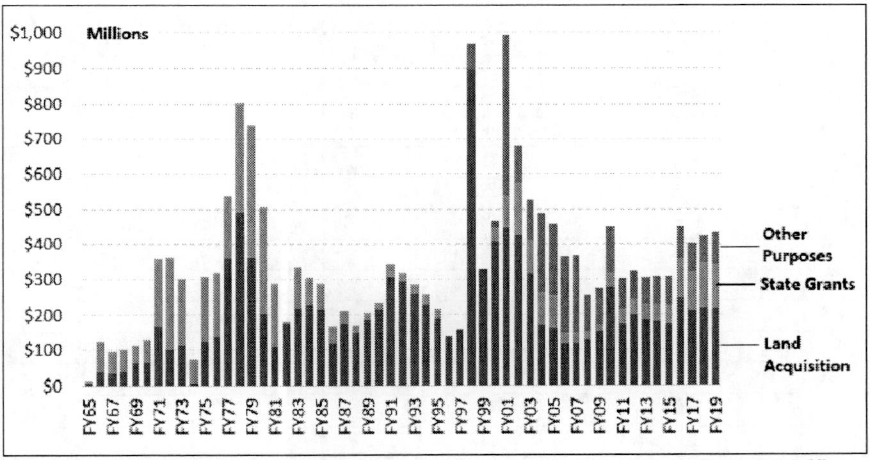

Source: Graphic created by CRS. The primary source for the data is the DOI Office of Budget. Additional sources of information include the annual DOI Budget in Brief and congressional documents accompanying the annual Interior appropriations bill.

Notes: The graph does not reflect $76 million provided for the transition quarter from July 1, 1976, to September 30, 1976, due to a change by Congress to the fiscal year calendar. Amounts shown represent the appropriations in the fiscal years indicated and thus are not adjusted for inflation.

Figure 4. LWCF Annual Discretionary Appropriations, FY1965-FY2019. (In millions of dollars, not adjusted for inflation)

[41] The LWCF had accumulated receipts sufficient to cover an appropriation exceeding the annual authorization. Specifically, in 1997 the LWCF had a balance of $11.9 billion in unappropriated receipts, which represented the difference between the receipts into the fund and the appropriations from the fund since its creation. Appropriations also exceeded the authorized level in FY2001.

Table 2. LWCF Annual discretionary appropriations by type, FY1965-FY2019 (in millions of dollars, not adjusted for inflation)

Fiscal Year	Land Acquisition	State Grants	Other Purposes	Total
1965	$6	$10	$0	$16
1966	38	84	0	122
1967	36	59	0	95
1968	40	64	0	104
1969	64	48	0	112
1970	66	65	0	131
1971	168	189	0	357
1972	102	259	0	362
1973	113	187	0	300
1974	5	71	0	76
1975	122	186	0	307
1976	136	181	0	317[a]
1977	356	182	0	538
1978	491	314	0	805
1979	361	376	0	737
1980	202	307	0	509
1981	108	180	0	289
1982	176	4	0	180
1983	220	115	0	335
1984	227	75	0	302
1985	213	74	0	287
1986	121	47	0	168
1987	176	35	0	211
1988	150	20	0	170
1989	186	20	0	206
1990	212	20	0	231
1991	308	33	0	342
1992	294	23	0	317
1993	256	28	0	284
1994	228	28	0	256
1995	189	28	0	217
1996	137	1	0	138
1997	158	1	0	159
1998	896	1	72	969
1999	328	0	0	328
2000	406	41	20	467
2001	449	90	456	995[b]
2002	429	144	105	677
2003	313	97	118	529
2004	170	94	225	488

Fiscal Year	Land Acquisition	State Grants	Other Purposes	Total
2005	164	91	203	459
2006	120	30	213	363
2007	120	30	216	366
2008	129	25	101	255
2009	152	19	104	275
2010	278	40	132	450
2011	177	40	84	301
2012	199	45	78	322
2013	187	43	74	303
2014	180	48	78	306
2015	178	48	80	306
2016	247	110	93	450
2017	209	110	81	400
2018	220	124	81	425
2019	218	124	93	435

Sources: The primary source for the data is the DOI Office of Budget. Additional sources of information include the annual DOI Budget in Brief and congressional documents accompanying the annual Interior appropriations bill.

Notes: Figures do not reflect $76 million provided for the transition quarter from July 1, 1976, to September 30, 1976. Also, dollars represent the appropriations in the fiscal years indicated and thus are not adjusted for inflation.

[a] The FY1976 total appropriation was approximately $393 million including the $76 million for the transition quarter.

[b] This figure includes approximately $153 million of Forest Service appropriations in law that were not "warranted" from LWCF, and thus were not officially recorded as taken from this fund. The Government Accountability Office defines warrant as "An official document that the Secretary of the Treasury issues upon enactment of an appropriation that establishes the amount of moneys authorized to be withdrawn from the central accounts that the Department of the Treasury maintains." See U.S. Government Accountability Office, A Glossary of Terms Used in the Federal Budget Process, September 2005, p. 101, http://www.gao.gov/new.items/d05734sp.pdf. Excluding the $153 million that was not warranted, the FY2001 LWCF total was approximately $842 million.

Another spike occurred in FY2001, when discretionary appropriations again exceeded the authorized level and totaled nearly $1 billion. This record level of funding was provided partly in response to President Clinton's Lands Legacy Initiative, which sought $1.4 billion for 21 resource protection programs, including the LWCF. It also was provided in response to some congressional interest in securing increased and more certain funding for the LWCF. The 106th Congress considered legislation to fully fund the LWCF and to make it operate like a private-sector trust fund. Such proposals sought to divert offshore oil and gas revenues to a Conservation and Reinvestment Act (CARA) Fund and to permanently appropriate receipts credited to the LWCF, among other related purposes. When it became clear that CARA legislation would not be enacted, Congress included aspects of the legislation in the FY2001 Interior and Related Agencies Appropriations law (P.L. 106-291).[42]

Over the last 10 years (FY2010-FY2019), LWCF discretionary appropriations began with a period high of $450.4 million in FY2010, then declined to roughly $300 million to $322 million through FY2015. In contrast, since FY2016, the appropriations have been between $400 million and $450 million, and they ended the period with $435.0 million for FY2019. The FY2010 high was about half the authorized level, and the $435.0 million for FY2019 was nearly half the authorized level. (See *Table 3* for annual discretionary appropriations, and the portions for acquisition of each agency, the stateside program, and other purposes, over the last 10 years.)

The land acquisition total ranged from a low of $176.8 million in FY2011 to a high of $277.9 million in FY2010. In addition to FY2010, the portion dedicated to land acquisition for all agencies was at least $200 million for the four most recent fiscal years (FY2016-FY2019), and most recently it was $217.7 million in FY2019.

During this 10-year period, appropriations for the stateside program fluctuated between $39.9 million for FY2011 and $124.0 million for each of FY2018 and FY2019. The appropriation for each of the first six fiscal years (FY2010-FY2015) ranged between $40 and $48 million. In contrast, the appropriation for each of the last four fiscal years (FY2016-FY2019) was between $110 and $124 million. Moreover, the appropriation for each of the last two years (FY2018- FY2019) was more than triple the appropriation for each of the first two years (FY2010-FY2011).

---

[42] These provisions established the Conservation Spending Category (CSC), with the LWCF as a major component in the CSC. The CSC provisions set a target for total funding for all the component programs in FY2001 at $1.6 billion, including $1.2 billion through Interior appropriations and $400 million through Department of Commerce appropriations. Under law, the target was to increase each year until it reached $2.4 billion in FY2006. However, Congress generally did not use the CSC structure in appropriating funds to the LWCF and related programs.

Table 3. LWCF Annual discretionary appropriations by type and agency, FY2009-FY2019 (in millions of dollars, not adjusted for inflation)

Purpose	FY 2010	FY 2011	FY 2012	FY 2013	FY 2014	FY 2015	FY 2016	FY 2017	FY 2018	FY 2019
Land Acquisition										
Bureau of Land Management	$29.7	$22.0	$22.3	$21.2	$19.5	$19.7	$38.6	$31.4	$24.9	$26.5
Fish and Wildlife Service	86.3	54.9	54.6	50.8	54.4	47.5	68.5	60.0	63.8	65.2
National Park Service	86.3	54.9	57.0	52.8	50.0	50.8	63.7	52.0	56.9	44.4
Forest Service	63.5	32.9	52.5	49.8	43.5	47.5	63.4	54.4	64.3	72.6
DOI OVS[a]	12.1	12.1	12.7	12.0	12.2	12.0	12.6	11.0	10.2	9.0
Total Land Acquisition	*277.9*	*176.8*	*199.1*	*186.6*	*179.6*	*177.6*	*246.9*	*208.8*	*220.3*	*217.7*
State Grants	40.0	39.9	44.9	42.6	48.1	48.1	110.0	110.0	124.0	124.0
Other Purposes	132.5	83.8	78.3	74.2	78.4	80.4	93.1	81.1	80.7	93.3
Total	$450.4	$300.5	$322.3	$303.3	$306.0	$306.1	$450.0	$400.0	$425.0	$435.0

Sources: The primary source for the data is the DOI Office of Budget. Additional sources of information include the annual DOI Budget in Brief and congressional documents accompanying the annual Interior appropriations bill.

Notes: Dollars represent the appropriations in the fiscal years indicated and thus are not adjusted for inflation. Figures reflect any reductions due to the use of prior-year funds.

[a] OVS is the Office of Valuation Services. Figures reflect appropriations from LWCF to DOI Departmental Management for land acquisition appraisal services.

The appropriations for other purposes were at their highest in the first year of the period, with $132.5 million in FY2010. Thereafter they did not exceed $100 million. They ranged between $74 million and $93 million, ending with $93.3 million in FY2019, the second highest amount during the 10-year period.

Both the dollar amount and percentage of LWCF appropriations provided to other purposes have varied widely since FY1998, as shown in *Table 4*. The dollar value of the appropriations for other purposes was much higher in FY2001 than any other year, when these appropriations were used to fund programs in the Clinton Administration's Lands Legacy Initiative.

Table 4. LWCF discretionary appropriations for other purposes, FY1998-FY2019 (in millions of dollars, not adjusted for inflation)

Fiscal Year	LWCF Total Discretionary Appropriations	Appropriations for Other Purposes	Other Purposes as % of Discretionary Appropriations
FY1998	$969.1	$72.0	7%
FY1999	328.2	0	0
FY2000	466.9	20.0	4
FY2001	995.4[a]	456.0	46
FY2002	677.2	104.6	15
FY2003	528.9	118.4	22
FY2004	488.1	224.7	46
FY2005	459.0	203.5	44
FY2006	362.8	213.1	59
FY2007	366.1	216.1	59
FY2008	255.1	101.3	40
FY2009	275.3	104.1[b]	38
FY2010	450.4	132.5	29
FY2011	300.5	83.8	28
FY2012	322.3	78.3	24
FY2013	303.3	74.2	24
FY2014	306.0	78.4	26
FY2015	306.1	80.4	26
FY2016	450.0	93.1	21
FY2017	400.0	81.1	20
FY2018	425.0	80.7[c]	19
FY2019	435.0	93.3	21

Sources: The primary source for the data is the DOI Office of Budget. Additional sources of information include the annual DOI Budget in Brief and congressional documents accompanying the annual Interior appropriations bill.

Notes: Dollars represent the appropriations in the fiscal years indicated and thus are not adjusted for inflation.

[a] This figure includes approximately $153 million of Forest Service appropriations in law that were not "warranted" from LWCF.

[b] This figure has been reduced by $8.0 million due to the use of prior-year funds.

[c] This figure has been reduced by $5.9 million due to the use of prior-year funds.

The highest percentage of funds provided for other purposes (59%) occurred in FY2006 and FY2007, in response to President George W. Bush's request for funding for an array of other programs. In some years, Congress has appropriated significantly less for other purposes than the Administration has requested. For instance, for FY2008 the Bush

Administration sought $313.1 million for other purposes of a total request of $378.7 million. Congress appropriated $101.3 million for other purposes of a total appropriation of $255.1 million.

ISSUES FOR CONGRESS

Through authorizing and appropriations legislation, hearings, and other debates, Congress has considered an array of issues related to the LWCF. Several of these issues are discussed below, including those that were addressed in provisions of recently enacted legislation—P.L. 116-9.

Changes Enacted in P.L. 116-9

Permanent Reauthorization of Revenue Provisions

The 116th and other recent Congresses debated whether to extend provisions of the LWCF Act that allowed $900 million in revenues to be deposited annually in the LWCF (through a particular fiscal year) and, if so, whether to authorize a short, long, or permanent extension. In 2019, the 116th Congress made these revenue deposits permanent, as noted above.[43]

The revenue deposit provisions originally contained long-term expiration dates of roughly 25 years and later contained a short-term expiration of 3 years. As enacted in 1964, the LWCF Act had established June 30, 1989, as the general end date for revenues to accrue to the LWCF; in 1976, this date was changed to *September* 30, 1989.[44] In 1987, the general end date for revenues to accrue to the fund was changed from September 30, 1989, to September 30, 2015.[45] Later in 2015, after the

---

[43] This change was enacted in P.L. 116-9 (§3001) on March 12, 2019, as codified at 54 U.S.C. §200302.

[44] The LWCF Act (P.L. 88-578) had established June 30, 1989, as the general end date for revenues to accrue to the LWCF. In 1976, the date of *June* 30, 1989, was changed to *September* 30, 1989, apparently to reflect the new start of the fiscal year (on October 1).

[45] §5201(f), P.L. 100-203, the Omnibus Budget Reconciliation Act of 1987, enacted December 22, 1987.

expiration of this provision, the general end date for revenues to accrue to the fund was changed from September 30, 2015, to September 30, 2018.[46] Following its expiration, this provision was made permanent in 2019.

In favor of a permanent or relatively long-term extension, LWCF advocates asserted that the program has successfully protected valuable lands and resources across the nation for more than 50 years and provided lands and infrastructure needed for public health, recreation, and tourism. They contended that continuity and stability of the land protection and recreational opportunities that LWCF supports are important national goals that will help communities attract businesses, new residents, and tourists. They noted the sufficiency of LWCF's dedicated funding stream for supporting the program.

Other Members and stakeholders proposed a short-term extension or expiration of the revenue provisions, on the grounds that LWCF had sufficient funds to meet the nation's land acquisition and recreation needs for many years to come. They pointed to the relatively large balance of unappropriated monies in the LWCF that Congress could appropriate, as well as to the substantial levels of mandatory funding that would be derived for the stateside program under GOMESA. Some advocated for structural reforms of the program before a long-term extension could be viable, such as reforms to direct funding away from federal land acquisition and to other activities, such as maintenance of federal lands and facilities or programs that support states. Still others favored depositing the OCS revenues in the General Fund of the Treasury (rather than in the LWCF) to support federal activities generally or specific priority programs.

Specify Stateside Share of Funding

The 116th and earlier Congresses considered whether to amend the LWCF Act to specify a portion of the appropriations for the stateside program and, if so, what amount would be optimal. As noted, in 2019, the 116th Congress amended the LWCF Act to specify funding for "financial assistance to states" of "not less than 40%" of the total made available

[46] Title VIII, Division O, P.L. 114-113, the Consolidated Appropriations Act, 2016, enacted December 18, 2015.

through appropriations or deposits under GOMESA.[47] Provisions of the LWCF Act on the share of funding for state and federal activities have changed throughout the act's history.

The level of discretionary appropriations for the stateside program likely will be of continued interest in part due to the substantial increase in mandatory funding under GOMESA, beginning in FY2018. In some past years, Congress and/or the Administration did not support funds, or supported relatively low levels of funds, for new stateside grants. Cited reasons have included that state and local governments have alternative sources of funding for parkland acquisition and development, the current program could not adequately measure performance or demonstrate results, the federal government supports states through other natural resource programs, and large federal deficits require a focus on core federal responsibilities.

Stateside supporters contend that the program contributes significantly to statewide recreation planning, state leadership in protection and development of recreation resources, and long-term outdoor recreation in areas that are readily accessible to communities. They see the program as a way to help fiscally constrained local governments and leverage state and local funds for recreation. Further, advocates assert that investments in recreation save money in other areas; for instance, they say that these investments promote healthier lifestyles and thus save health care expenditures.[48] Some prefer state acquisition (and development) of lands with LWCF funds over federal acquisition of lands.

Specify Recreational Access Funding

The 116th Congress, as well as previous Congresses, discussed whether to amend the LWCF Act to set aside a portion of federal acquisition

---

[47] This change was enacted in P.L. 116-9 (§3001) on March 12, 2019, as codified at 54 U.S.C. §200304(b).

[48] As an example of testimony supporting increased funding for the stateside program, see Prepared Statement of Administrator, Wyoming State Parks Administrator and Vice President of the National Association of State Park Directors Domenic Bravo, in U.S. Congress, House Committee on Natural Resources, Subcommittee on Federal Lands, *Federal Land Acquisition and its Impacts on Communities and the Environment*, hearing, 114th Cong., 1st sess., April 15, 2015, pp. 1-2.

funding to facilitate access to federal lands for recreational purposes. As noted earlier, in 2019, the 116th Congress amended the law to set an annual threshold for such acquisitions—not less than $15.0 million or 3% of amounts made available for expenditure (under 54 U.S.C. §200303), whichever is greater.[49]

Most federal lands are accessible to the public, but some areas have limited access or are unavailable to the public. This is sometimes the case because private or state lands, or geographical features such as mountains or rivers, limit or block public access. Proponents of enhanced recreational access sought to amend the LWCF Act to specify funding for related federal acquisitions. The focus was on acquisition of lands and interests (e.g., easements and rights of way) that support access for hunting and fishing, among other recreational pursuits. Proponents cited insufficient opportunity for recreating in some areas and broader health, economic, and other values of recreating on public lands.

Opposition to such a set-aside may have been related to concerns about the extent of federal land ownership, preferences for prioritizing LWCF for other purposes, or satisfaction that the agencies are sufficiently prioritizing recreational access. According to BLM testimony in 2015, for instance, "nearly 100 percent of LWCF funding over the past several years has been used for projects that enhance public access for recreation."[50] For the Forest Service, for example, 39 of the 40 acquisitions completed in FY2014 with LWCF funding provided improved access or legal access where none existed, according to agency testimony in 2015.[51] However,

---

[49] This change was enacted in P.L. 116-9 (§3001) on March 12, 2019, as codified at 54 U.S.C. §200306(c).

[50] Statement of Bureau of Land Management's Deputy Director for Operations Steve Ellis, in U.S. Congress, Senate Committee on Energy and Natural Resources, *Legislative Hearing on S. 556, the Bipartisan Sportsmen's Act of 2015*, hearing, 114th Cong., 1st sess., March 12, 2015, p. 6, http://www.energy.senate.gov/public/index.cfm/hearings-andbusiness-meetings?ID=f46d20ea-6d78-4c4a-a582-2881bbb5abca.

[51] Testimony of the Forest Service's Deputy Chief for National Forest System Leslie Weldon, in U.S. Congress, Senate Committee on Energy and Natural Resources, *Legislative Hearing on S. 556, the Bipartisan Sportsmen's Act of 2015*, hearing, 114th Cong., 1st sess., March 12, 2015, pp. 5-6, http://www.energy.senate.gov/public/index.cfm/hearings-andbusiness-meetings?ID=f46d20ea-6d78-4c4a-a582-2881bbb5abca.

the extent to which agencies prioritize acquisition funding for recreational access might vary among all agencies and from year to year.

Current Issues

Level of Funding

There continue to be differing opinions as to the optimal level of LWCF appropriations. The LWCF has broad support from resource protection advocates, many of whom seek stable and predictable funding through consistent levels of appropriations. Most of these advocates seek higher appropriations in general. For instance, for several years the Obama Administration proposed appropriations of $900 million for LWCF. Some advocates have specific priorities, such as higher acquisition funding for one of the four federal agencies, the state grant program, or a particular site or area. Advocates of higher federal land acquisition funding promote a strong federal role in acquiring and managing sensitive areas and natural resources.

In contrast, there is broad opposition to the LWCF based on varied concerns, with opponents generally seeking reduced levels of funds for LWCF. Some opponents seek to reduce the size of the federal estate and minimize further federal acquisition of land either generally or at specific sites, especially in the West, where federal ownership is already concentrated. Other concerns involve preferences for private land ownership or state and local ownership and management of resources, potential impacts of federal land ownership on uses of private lands, and reduced local tax revenues that result from public ownership. For instance, in the past, the House Committee on the Budget supported reducing the federal estate and giving state and local governments more control of resources within their boundaries.[52] Others have sought LWCF reductions as part of a broader focus on reducing the federal deficit.

---

[52] House Committee on the Budget, *Concurrent Resolution on the Budget—Fiscal Year 2016*, H.Rept. 114-47 on H.Con.Res. 27, p. 119. The committee's reports on budget resolutions for subsequent fiscal years (FY2017-FY2019) expressed a similar view.

Mandatory vs. Discretionary Appropriations

Some advocates of higher LWCF funding seek partial or full permanent appropriations of monies under the LWCF Act. For instance, the Obama Administration proposed $900 million for LWCF for FY2017 through a combination of discretionary ($475.0 million) and mandatory ($425.0 million) appropriations. Further, the Obama Administration proposed amending the LWCF Act to appropriate mandatory funding of $900 million annually beginning in FY2018.[53] As attributes of the mandatory approach, advocates cite the certainty and predictability of program funding generally and in particular in fostering the ability of agencies to undertake multiyear, large-scale, and collaborative acquisitions.[54] They also note the existence of a dedicated funding stream for LWCF at the authorized level of $900 million and the original intent of Congress that revenues to the fund be used for the LWCF Act's purposes. Questions have included how to offset any new permanent appropriations and how to allocate permanent appropriations among different LWCF programs and purposes.

Proponents of discretionary appropriations support retaining Congress's authority to determine annually the level of funding needed for LWCF overall and for the various program components. They note that through the discretionary process, the need for LWCF appropriations can be assessed in comparison with other natural resource programs and broader governmental needs. This approach also has been valued as providing recurring opportunities for program oversight. For instance, in the past, the House Committee on the Budget supported keeping land acquisition funding as discretionary to allow for regular oversight.[55]

---

[53] U.S. Dept. of the Interior, *Fiscal Year 2017: The Interior Budget in Brief*, p. DH-68, at https://www.doi.gov/budget. Proposals to establish mandatory funding for the LWCF also have been introduced in the 116th and prior Congresses.

[54] Testimony of Deputy Secretary of the Interior Michael Connor, in U.S. Congress, Senate Committee on Energy and Natural Resources, *Hearing on the Reauthorization and Potential Reforms to the Land and Water Conservation Fund (LWCF)*, hearing, 114th Cong., 1st sess., April 22, 2015, http://www.energy.senate.gov/public/index.cfm/hearings-andbusiness-meetings?ID=9a8c795c-f8fe-4168-af7c-1a89ff5a3fee.

[55] House Committee on the Budget, *Concurrent Resolution on the Budget—Fiscal Year 2019*, H.Rept. 115-816 on H.Con.Res. 128, p. 46. The committee's reports on budget resolutions for earlier fiscal years (e.g., FY2016-FY2018) expressed a similar view.

Maintenance vs. Acquisition

There are differing opinions as to how LWCF funds should be used. Some Members and others have asserted that maintaining (and rehabilitating) the land and facilities that federal agencies already own should take priority over further acquisitions. For instance, in an opening statement in 2015, the chair of the Senate Committee on Energy and Natural Resources expressed interest in shifting "the federal focus away from land acquisition, particularly in Western states, toward maintaining and enhancing the accessibility and quality of the resources that we have."[56] Some cite the deferred maintenance needs of the agencies as a higher priority than acquisition, and one that requires funding through additional sources, including LWCF.[57] As an example, the Trump Administration did not initially seek funding for individual acquisitions by agencies in FY2019, asserting that at a time when DOI has "billions of dollars in deferred maintenance, land acquisitions are lower priority activities than maintaining ongoing operations and maintenance."[58]

Supporters of retaining a strong LWCF role in acquisition cite a continued need to preserve iconic resources and provide additional opportunities for public recreation and other land uses. They assert economic benefits of federal land ownership and improvements of federal land management through consolidation of ownership.[59] In addition, since federal agencies cannot use LWCF funds for maintenance, some supporters of this priority favor more funding to other accounts that can be used for

---

[56] Statement of Senator Lisa Murkowski, in U.S. Congress, Senate Committee on Energy and Natural Resources, *Hearing on the Reauthorization and Potential Reforms to the Land and Water Conservation Fund (LWCF)*, hearing, 114th Cong., 1st sess., April 22, 2015, http://www.energy.senate.gov/public/index.cfm/hearings-and-business-meetings?ID=9a8c795c-f8fe-4168-af7c-1a89ff5a3fee.

[57] For information on the deferred maintenance estimates of the land management agencies over 10 years, see CRS Report R43997, *Deferred Maintenance of Federal Land Management Agencies: FY2009-FY2018 Estimates and Issues*, by Carol Hardy Vincent. For details on NPS deferred maintenance, see CRS Report R44924, *The National Park Service's Maintenance Backlog: Frequently Asked Questions*, by Laura B. Comay.

[58] Office of Management and Budget, An American Budget: Major Savings and Reforms, Fiscal Year 2019, p. 56.

[59] See, for example, Host of Fresh Tracks and On Your Own Adventures Randy Newberg, in U.S. Congress, House Committee on Natural Resources, Subcommittee on Federal Lands, Federal Land Acquisition and its Impacts on Communities and the Environment, hearing, 114th Cong., 1st sess., April 15, 2015, pp. 2-4.

maintenance while retaining the LWCF for acquisitions.[60] Such accounts are within the land management agency budgets as well as within the Department of Transportation (for roads).

Stateside Program

Whether to change the apportionment of funds under the stateside program has been debated. Under the traditional state grant program, a portion of the appropriation is to be distributed equally among the states. The percentage varies depending on the total amount of appropriations, but it has been 40% in recent years.[61] Further, under the LWCF Act, the Secretary of the Interior has discretion to apportion the balance based on need, and population has been the biggest factor in determining need. Questions include whether to alter the portion of funding that is distributed equally among states and the factors for allocating the balance.

Since FY2014, Congress has approved a portion of the state grant funds for a competitive grant program. The extent to which the state grant program should be competitive is of ongoing interest. Questions include whether to establish a competitive program in law, which criteria should be used in any competitive program, and whether to authorize mandatory appropriations under GOMESA to be used for a competitive program. Currently, the GOMESA monies are used only for the traditional grant program.[62]

Other Programs

Congress is examining whether purposes other than federal land acquisition and stateside grants should be authorized in the LWCF Act and appropriated from the fund. As discussed above, whether to authorize the fund for maintenance is an area of focus. Some Presidents and Members of Congress have sought to channel LWCF funding to an array of land

---

[60] Statement of Senator Maria Cantwell, in U.S. Congress, Senate Committee on Energy and Natural Resources, *Hearing on the Reauthorization and Potential Reforms to the Land and Water Conservation Fund (LWCF)*, hearing, 114th Cong., 1st sess., April 22, 2015, p. 3, http://www.energy.senate.gov/public/index.cfm/hearings-and-businessmeetings?ID=9a8c795c-f8fe-4168-af7c-1a89ff5a3fee.

[61] See 54 U.S.C. §200305(b)(1).

[62] P.L. 109-432, Division C, §105(a)(2)(B).

protection purposes. For instance, the George W. Bush Administration requested LWCF funds for cooperative conservation programs through which federal land managers partner with other landowners to protect natural resources and improve recreation on lands under diverse ownership. The Obama Administration also supported the use of LWCF funds for other purposes, although generally fewer than the Bush Administration.[63] Some proposals seek to enhance funding for programs that benefit states, such as grants under the Cooperative Endangered Species Conservation Fund and the Forest Legacy Program, among others. These proposals could shift resources from federal lands to state grant programs.

A factor in the debate has been the unappropriated balance in the fund and whether to allow these funds to be used for broader purposes than those currently authorized. Traditional fund beneficiaries and others have expressed concern about expanding the uses of appropriations, particularly if that expansion is accompanied by reductions in the amount available for federal land acquisition or state grants. Proponents of funding other programs assert that the sizeable unappropriated balance in the LWCF would be sufficient to cover the traditional purposes under the LWCF Act as well as additional ones.

---

[63] Past legislative proposals also would have provided LWCF funding to a more diverse programs and activities. As examples from the 115th Congress, H.R. 343 would have authorized the Secretary of Housing and Urban Development to use LWCF funding to provide financial assistance to entities to carry out park and infrastructure projects. H.R. 5170 would have authorized LWCF funding for deferred maintenance, critical infrastructure, visitor services, and clean-up efforts on FS, NPS, FWS, and BLM lands; the Payments in Lieu of Taxes Program; and promotion of off-shore energy exploration, innovation, and education, among other activities.

In: Key Congressional Reports for June 2019 ISBN: 978-1-53616-662-0
Editor: Stephanie A. Betts

Chapter 2

VEHICLE ELECTRIFICATION: FEDERAL AND STATE ISSUES AFFECTING DEPLOYMENT*

Bill Canis, Corrie E. Clark and Molly F. Sherlock

SUMMARY

Most of the 270 million cars, trucks, and buses on U.S. highways are powered by internal combustion engines using gasoline or diesel fuel. However, improvements in technology have led to the emergence of vehicle electrification as a potentially viable alternative to internal combustion engines. Several bills pending in the 116th Congress address issues and incentives related to electric vehicles and charging infrastructure.

Experience with fully electric vehicles is relatively recent: While a few experimental vehicles were marketed in the United States in the 1990s, the first contemporary all-electric passenger vehicles were introduced in 2010. Since then, newer models have increased the range an electric vehicle can travel on a single charge, and charging stations have

* This is an edited, reformatted and augmented version of Congressional Research Service Publication No. R45747, dated June 3, 2019.

become more readily available. These developments have been spurred by a range of government incentives, both in the United States and abroad. Transit buses are the fastest-growing segment of vehicle electrification in China, while in the United States and the European Union, the pace of bus electrification is slower.

In the United States, federal incentives for electric passenger vehicle purchases have remained largely unchanged for more than a decade and are based primarily on tax credits for electric vehicle purchases and recharging infrastructure investments, and spending on battery chemistry research to develop less-expensive technologies:

- The plug-in electric tax credit permits a taxpayer to take a credit of up to $7,500 for each vehicle that can be recharged from the electricity grid; it phases out after a manufacturer has sold 200,000 eligible vehicles, a threshold that has been met by Tesla and General Motors.
- A tax credit for installation of alternative fuel vehicle refueling property expired in 2017; it had allowed a tax credit of $1,000 for equipment installed at a residence and up to $30,000 for business installations.
- Investment in transportation electrification research and development (R&D), which has led to the gradual reduction in the cost of producing lithium-ion batteries, is administered by the U.S. Department of Energy (DOE) in cooperation with private industry. Although the Trump Administration has recommended large reductions in these programs, Congress has maintained annual funding for sustainable transportation of nearly $700 million in the past two fiscal years.

Other programs that directly influence the level of vehicle electrification include the DOE Clean Cities Program, which supports local efforts to reduce fossil fuel-powered transportation, and the Department of Transportation's Alternative Fuel Corridors, which are designated Interstate Highway corridors with a sufficient number of alternative fueling stations, including electric vehicle chargers, to allow alternative fuel vehicles to travel long distances. The federal government also funds municipal transit bus electrification through Federal Transit Administration grants, which may be used for the purchase of all-electric buses. The pace of electrification also may be affected by proposals for less stringent federal standards for Corporate Average Fuel Economy (CAFE) and greenhouse gas emissions from vehicles.

Beyond these federal programs, states and electric utilities provide a range of incentives for electrification. The National Conference of State Legislatures reports that 45 states and the District of Columbia offer incentives such as income tax credits for electric vehicle and charger purchases, reduced registration fees, and permitting solo drivers of electric vehicles to use carpool lanes. The California Zero Emission Vehicle program is spurring sales of electric vehicles in 10 states. Utilities can provide incentives to charge during off-peak hours, install public electric charging infrastructure, and utilize vehicle-togrid (V2G) storage. V2G storage would allow idle vehicle batteries to supply electricity to the grid rather than drawing power from it during peak demand periods.

INTRODUCTION

Motor vehicle electrification has emerged in the past decade as a potentially viable alternative to internal combustion engines. Although only a small proportion of the current motor vehicle fleet is electrified, interest in passenger vehicle electrification has accelerated in several major industrial countries, including the United States, parts of Europe, and China. Despite advances in technology, electric vehicles (EVs) continue to be significantly more expensive than similarly sized vehicles with internal combustion engines. For this reason, governments in many countries have adopted policies to promote development and sales of electric vehicles. This chapter discusses federal and state government policies in the United States to support electrification of light vehicles and transit buses, as well as proposals to reduce or eliminate such support.

BACKGROUND ON MOTOR VEHICLE ELECTRIFICATION

More than 92 million light vehicles—passenger cars, pickup trucks, and SUVs—were sold worldwide in 2018. The three largest markets were China (27 million vehicles sold), Europe (20 million), and the United States (17 million).[1] Most of these vehicles are powered by internal combustion engines.

---

[1] Ward's Database, "China, Europe and U.S. Vehicle Sales 2018." If Canada and Mexico are included, then the North American market would tie with Europe as the second largest.

The global market for electrified vehicles is small but growing: In 2018, more than 2 million plug-in hybrid and battery electric vehicles were sold worldwide, a 64% increase over 2017.[2] These account for about 2% of all passenger vehicle sales, both worldwide and in the United States. Demand for electric vehicles is expected to continue to grow, as some countries have called for a complete shift away from sales of new fossil-fuel vehicles by 2030.[3]

The market for urban transit buses is smaller than the passenger car and SUV markets, but electric vehicles make up a larger part of its footprint. China leads in this category, with 106,000 electric buses put in service in 2017, bringing its total electric bus fleet to 384,000. It has been forecast to remain the largest electric bus market going forward. In the European Union (EU) and the United States, the pace of electrification is slower: More than 200 electric buses were sold in the EU in 2017, bringing the total in service to 1,700; in the United States, approximately 100 electric buses were sold, bringing the total to 300.[4]

Two basic types of electric vehicles are now in use:

- Hybrid electric vehicles (HEVs) have both internal combustion engines and electric motors that store energy in batteries. They do not plug into external sources of electricity, but use regenerative braking and the internal combustion engine to recharge.
- Plug-in electric vehicles, of which there are two types: plug-in hybrid electric vehicles (PHEVs) use an electric motor and an internal combustion engine for power, and they use electricity

---

[2] InsideEVs, "Worldwide Sales of Electric Vehicles," viewed March 5, 2018, at https://insideevs.com/monthly-plug-inev-sales-scorecard-historical-charts/.

[3] At least eight countries have set goals of phasing out sales of new internal combustion vehicles by 2030, although apparently none of them has taken binding legislative action. As of September 2018, the following countries have called for no sales of new internal combustion engine vehicles by 2030: Austria, Costa Rica, Germany, India, Ireland, Israel, the Netherlands, and Norway. In addition, France and the United Kingdom have set similar goals for 2040. Isabella Burch and Jock Gilchrist, *Survey of Global Activity to Phase Out Internal Combustion Engine Vehicles*, Center for Climate Protection, September 2018, at https://climateprotection.org/our-work/reports.

[4] Aleksandra O'Donovan, *The Global Electric Bus Market Gets Into Gear (Part 2)*, Bloomberg New Energy Finance, February 1, 2018, pp. 12-14.

from an external source to recharge the batteries. Battery electric vehicles (BEVs) use only batteries to power the motor and use electricity from an external source for recharging.[5] In this chapter, electric vehicles refer to these two types of plug-in vehicles, unless otherwise noted.

Electrification of vehicles has been limited by three factors: (1) the high cost of producing the lithium ion batteries (currently the preferred battery chemistry) that propel them; (2) their limited range; and (3) vehicle charging time and location. Not all motorists have easy access to charging stations at home or at work, and it can take several hours to fully charge the battery that powers the vehicle, depending on the type of charger used.

U.S. Trends

In 2018, more than 361,000 plug-in electric passenger vehicles (including PHEVs and BEVs) were sold in the United States, as well as more than 341,000 hybrid electric vehicles. This was the first year in which total sales of plug-in vehicles exceeded sales of hybrids (Figure 1). Nearly all automakers offer electric vehicles for sale: 42 different models were sold in 2018, with Tesla and Toyota recording the largest number of vehicle sales.[6] Sales of plug-in hybrid and battery electric vehicles in 2018 rose by over 80% from the previous year, bringing the total sales of plug-in vehicles since 2010 to just over 1 million.[7] The plug-in hybrid and battery electric share of the U.S. light vehicle market in 2018 was 2.1%.[8]

---

[5] U.S. Department of Energy, Alternative Fuels Data Center, at https://www.afdc.energy.gov/vehicles/electric_basics_ev.html.

[6] CRS analysis shows that of the 42 models sold domestically, 10 were produced at seven U.S. plants, with the remainder imported.

[7] U.S. Department of Energy, "One Million Plug-in Vehicles Have Been Sold in the United States," press release, November 26, 2018, at https://www.energy.gov/eere/vehicles/articles/fotw-1057-november-26-2018-one-million-plugvehicles-have-been-sold-united.

[8] CRS calculations based on Oak Ridge National Laboratory data; Oak Ridge National Laboratory, *Transportation Energy Data Book*, Table 6.2, at https://cta.ornl.gov/data/chapter6.shtml.

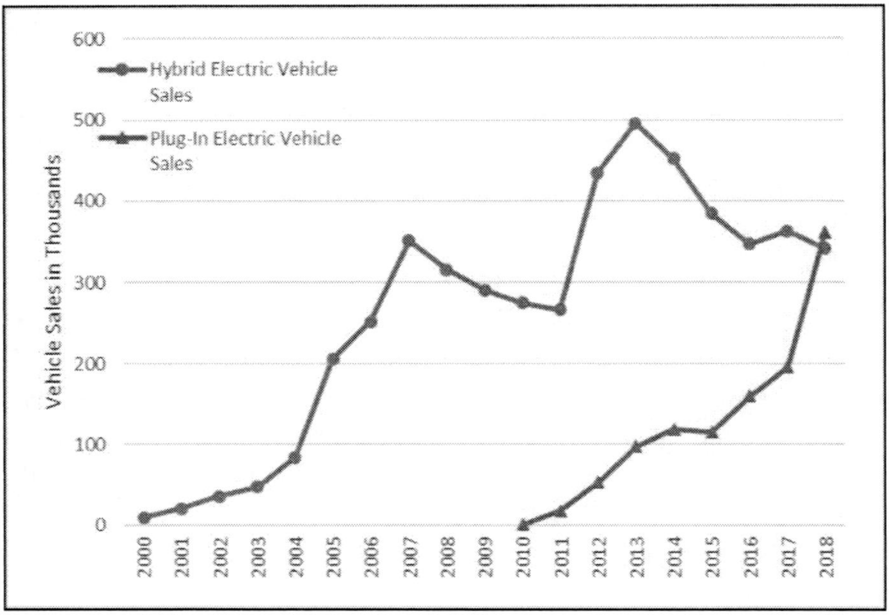

Source: Oak Ridge National Laboratory (ORNL), Transportation Energy Data Book, Table 6.2, at https://cta.ornl.gov/data/chapter6.shtml. 2018 data based on CRS calculations of unpublished ORNL data.

Notes: The total for plug-in electric vehicles includes plug-in hybrids and battery electric vehicles. Plug-in vehicle sales include only those vehicles certified for highway use.

Figure 1. U.S. Annual Hybrid Electric Vehicle and Plug-In Electric Vehicle Sales. (2000-2018)

The price of new electric vehicles is one factor inhibiting faster adoption. For example, the Leaf, a battery electric vehicle produced by Nissan, has a manufacturer's suggested retail price (MSRP) of $29,990, whereas the Nissan Sentra, a conventional vehicle similar in size and specifications to the Leaf, has an MSRP of $17,990. A smaller, less powerful vehicle with an internal combustion engine, the Nissan Versa, has an MSRP of $12,360; no electric counterpart is available in this price range. Electric vehicles are generally more expensive because of the high cost of producing the lithium-ion batteries that power them.[9]

[9] Similar pricing differences exist with other manufacturers: The manufacturer's suggested retail price (MSRP) for a model year 2019 Chevrolet Cruze is $17,995. The MSRP for a model

The federal government has supported vehicle electrification in several ways. There have been tax incentives for the purchase of vehicles as well as for construction of vehicle infrastructure, such as charging stations. Federal research and development investments have sought to reduce battery costs, increase vehicle range, and decrease charging times. The federal government has also made other investments to build out EV infrastructure.

FEDERAL TAX INCENTIVES FOR ELECTRIFICATION

Two types of tax incentives have been used to promote electric vehicles: consumer incentives for the purchase of plug-in electric vehicles and individual and business incentives to install electric-vehicle charging stations to expand the charging network.

Tax Credits for Vehicle Purchases

The credit for plug-in electric vehicles (Internal Revenue Code [IRC] §30D) is the primary federal tax incentive for electric vehicles.[10] The credit ranges from $2,500 to $7,500 per vehicle, depending on the vehicle's battery capacity. The tax credit is not a function of the vehicle's price. Therefore, the subsidy amount is larger (as a percentage of a vehicle's price) for less-expensive vehicles. Generally, taxpayers claim tax credits for vehicle purchases. If the purchaser or lessee is a tax-exempt organization, the seller of the vehicle may be able to claim the credit.

The plug-in electric vehicle credit begins phasing out after a vehicle manufacturer has sold 200,000 qualifying vehicles for use in the United

---

year 2019 Chevrolet Bolt EV (not including a tax credit and including some additional standard features) is $36,620. The MSRP comparison does not take into account the difference in operating costs over the lifetime of the vehicles, which may favor electric vehicles.

[10] See also CRS In Focus IF11017, *The Plug-In Electric Vehicle Tax Credit*, by Molly F. Sherlock.

States. General Motors (GM) and Tesla have reached the 200,000-vehicle limit, and tax credits for their vehicles have begun to phase down.[11]

Empirical studies have found that tax incentives lead to increased EV purchases.[12] However, particularly for higher-income taxpayers, the tax credit may be claimed for purchases that would have occurred absent a federal tax incentive. Some studies have also found that incentives given closer to the point of sale, such as a rebate given at the time of purchase, are more effective in stimulating vehicle sales than tax credits.[13]

The Joint Committee on Taxation (JCT) projects that the plug-in electric vehicle credit will reduce federal tax revenues by $7.5 billion

---

[11] Internal Revenue Service, "First Plug-In Electric Vehicle Manufacturer Crosses 200,000 Sold Threshold; Tax Credit for Eligible Consumers Begins Phase Down on Jan. 1," press release, December 14, 2018, at https://www.irs.gov/ newsroom/first-plug-in-electric-vehicle-manufacturer-crosses-200000-sold-thresholdtax-credit-for-eligible-consumersbegins-phase-down-on-jan-1; and Internal Revenue Service, "Plug-In Electric Vehicle Manufacturer Crosses 200,000 Sold Threshold; Tax Credit for Eligible Consumers Begins Phase Down on April 1," press release, March 26, 2019, at https://www.irs.gov/newsroom/plug-in-electric-vehicle-manufacturer-crosses-200000-sold-threshold-tax-credit-foreligible-consumers-begins-phase-down-on-april-1. Tesla vehicles will be eligible for a $3,750 credit until June 30, 2019, and then for a $1,875 credit until the end of 2019. See Tesla, "Electric Vehicle & Solar Incentives," at https://www.tesla.com/support/incentives. GM electric vehicles were eligible for the full $7,500 credit until March 31, 2019, and will be eligible for a $3,750 credit until September 30, 2019, and a $1,875 credit until March 31, 2020. See Chevrolet, "Bolt EV," at https://www.chevrolet.com/electric/bolt-ev-electric-car.

[12] See, for example, Alan Jenn, Katalin Springel, and Anand R. Gopal, "Effectiveness of Electric Vehicle Incentives in the United States," *Energy Policy*, vol. 119 (2018), pp. 349-356. That study found that every $1,000 in tax credits or rebates is associated with a 2.6% increase in EV sales. See also Jianwei Xing, Benjamin Leard, and Shanjun Li, *What Does an Electric Vehicle Replace?*, National Bureau of Economic Research (NBER) Working Paper 25771, April 2019, at https://www.nber.org/papers/w25771. That study found that in 2014, the federal income tax credit for EVs led to a 28.8% increase in EV sales. For surveys of the literature, see Yan Zhou, Todd Legin, and Steven E. Plotkin, *Plug-In Electric Vehicle Policy Effectiveness: Literature Review*, Argonne National Laboratory, ANL/ESD-16/8, May 2016; and Scott Hardman, Amrit Chandan, Gil Tal, and Tom Turrentine, "The Effectiveness of Financial Purchase Incentives for Battery Electric Vehicles—A Review of the Evidence," *Renewable and Sustainable Energy Reviews*, vol. 80 (2017), pp. 1100-1111. Shanjun Li, Lang Tong, Jianwei Xing, and Yiyi Zhou, "The Market for Electric Vehicles: Indirect Network Effects and Policy Design," *Journal of the Association of Environmental and Resource Economists*, March 2017, pp. 89-133, found that between 2011 and 2013, the federal tax credits contributed to about 40% of EV sales.

[13] Easwaran Narassimhan and Caley Johnson, "The Role of Demand-Side Incentives and Charging Infrastructure on Plug-In Electric Vehicle Adoption," *Environmental Research Letters*, vol. 13 (2018), pp. 1-11. That study found that a 1% increase in tax incentives relative to a vehicle's price was associated with a 1.15% to 2.91% increase in purchases, depending on the model specification.

between FY2018 and FY2022.[14] Any extensions to or expansions of the credit could increase this amount. About half of the forgone revenue is from credits claimed on corporate tax returns.[15] Additionally, the tax credits tend to be claimed by higher-income taxpayers. For 2016, 78% of the claimants filed returns with adjusted gross income (AGI) of $100,000 or more, and such returns accounted for 83% of the amount claimed.[16] (By comparison, of all returns filed, about 17% have AGI above $100,000.)

Legislation has been introduced in the 116th Congress that would modify the plug-in electric vehicle tax credit. Some bills propose expanding the credit. For example, the Driving America Forward Act (S. 1094/H.R. 2256) would increase the per-manufacturer cap to 600,000 vehicles and modify the credit during the phase-out period. The Electric Credit Access Ready at Sale (Electric CARS) Act of 2019 (S. 993/H.R. 2042) would extend the credit through December 31, 2029, and would also allow the credit to be transferred to the financing entity.

Other proposals would eliminate the credit. The Fairness for Every Driver Act (S. 343/H.R. 1027) would eliminate the credit and impose federal highway user fees on alternative fuel vehicles.[17]

Tax Incentives for Infrastructure

The primary federal tax incentive for EV infrastructure has been the tax credit for alternative fuel vehicle refueling property (IRC §30C), which

---

[14] Joint Committee on Taxation, Estimates of Federal Tax Expenditures for Fiscal Years 2018-2022, JCX-81-18, October 4, 2018.

[15] When a corporation claims the tax credit, it does not necessarily receive the full benefit. Ultimately, the incidence of the tax credits—the distribution of the benefits—depends on the relative responsiveness of sellers and buyers to changes in price. Tax credits claimed by sellers can be passed forward to buyers via lower prices. Conversely, sellers can raise prices to capture a portion of tax credits given to buyers.

[16] Internal Revenue Service, SOI Tax Stats—Individual Statistical Tables by Size of Adjusted Gross Income, "Table 3.3: All Returns: Tax Liability, Tax Credits, and Tax Payments, by Size of Adjusted Gross Income, Tax Year 2016 (Filing Year 2017)," at https://www.irs.gov/statistics/soi-tax-stats-individual-statistical-tables-by-size-of-adjusted-grossincome.

[17] Electric vehicle users do not pay into the Highway Trust Fund, which is funded by fuel taxes. See "EVs and Federal Highway Taxes " for more information.

expired in 2017. The credit was generally 30% of the cost of qualified property, with the credit limited to $30,000 for businesses at each separate location and $1,000 for property installed at a taxpayer's primary residence. For property sold to a tax-exempt entity, such as a school or a hospital, the seller of the property may have been able to claim the credit. Qualifying property included electric charging infrastructure as well as other forms of clean-fuel refueling property.

The credit for alternative fuel vehicle refueling property has been a temporary tax credit since first enacted in 2005. The credit has been extended six times, often retroactively. The credit most recently expired at the end of 2017, but could be extended again.[18] The uncertainty surrounding temporary tax incentives that are often retroactively reinstated diminishes their effectiveness as an investment incentive.[19] The Electric CARS Act of 2019 (S. 993/H.R. 2042) would extend the credit through 2029.

Data are not available on how much of the revenue loss associated with this provision is for EV infrastructure. The most recent one-year extension of this incentive, for all types of alternative fuel refueling property,[20] was estimated to reduce federal tax revenues by $67 million over the 10-year budget window.[21] Making the credit permanent for all types of qualifying property would reduce federal revenue by an estimated $332 million between FY2018 and FY2027.[22]

The tax credits for EV charging infrastructure that expired at the end of 2017, if extended, could support additional investment in Level 2

---

[18] For more information, see CRS Report R45347, *Tax Provisions That Expired in 2017 ("Tax Extenders")*, by Molly F. Sherlock; and CRS Report R44990, *Energy Tax Provisions That Expired in 2017 ("Tax Extenders")*, by Molly F. Sherlock, Donald J. Marples, and Margot L. Crandall-Hollick.

[19] The credit was temporarily larger in 2009 and 2010. The American Recovery and Reinvestment Act of 2009 (P.L. 111-5) temporarily increased the credit amount to 50% of the cost of qualified property. In addition, maximum credit amounts were increased to $50,000 for business property and $2,000 for nonbusiness property.

[20] Applicable fuels were electricity, natural gas, liquefied petroleum gas, hydrogen, E85, or diesel fuel blends containing a minimum of 20% biodiesel.

[21] Joint Committee on Taxation, Estimated Budget Effects of the Revenue Provisions Contained in the "Bipartisan Budget Act of 2018," February 8, 2018, JCX-4-18.

[22] Joint Committee on Taxation, *Federal Tax Provisions Expired in 2017*, March 9, 2018, JCX-5-18.

charging infrastructure. Expanded access to Level 2 charging at homes and workplaces could be a cost-effective solution to building out EV infrastructure in the near term. However, if electric vehicles are to be widely used for long-distance trips, a network of direct-current fast charger (DCFC) infrastructure (Level 3) is likely necessary. The tax credit is relatively small compared to the cost of a DCFC charging station. The high cost of this infrastructure, even if tax credits are extended, may continue to pose a barrier, especially if utilization rates are low.

Types of Electric Chargers

Level 1 charging allows vehicles to be charged using 120-volt service, but is slow, with vehicles needing to charge for at least 10 hours to supply 80 miles of range. Often installed in car owners' garages for overnight home charging, they cost about $300, not including installation; however, a standard 120-volt electrical outlet in a residential garage can provide the same charge, and a special charger is not necessary.

Level 2 charging requires a special charging unit and access to 240-volt service, and is faster, with one to six hours of charging sufficient to supply 80 miles of range. Charging units cost about $600, not including installation, and $6,000 per unit for commercial public installations.

Level 3 direct-current fast chargers (DCFCs) use up to 500 volts of direct current to supply 80 miles of range within minutes. They can cost up to $300,000 per unit, including permitting, design, and installation.

Given the differences in the costs and benefits associated with Level 2 and DCFC chargers, tax incentives could be provided that reflect some of these differences. There are few DCFC public chargers, yet having access to such infrastructure may have strong network benefits. Broadly, access to charging infrastructure has been shown in some studies to be a driver of demand for EVs.[23] If Congress wanted to encourage greater EV use, tax

[23] See Easwaran Narassimhan and Caley Johnson, "The Role of Demand-Side Incentives and Charging Infrastructure on Plug-In Electric Vehicle Adoption," *Environmental Research Letters*, vol. 13 (2018), pp. 1-11. This study found that Nissan Leaf purchases were highly correlated with access to charging infrastructure, measured as public charging stations per

credits could be designed to provide a larger incentive for investments in public DCFC infrastructure, relative to Level 2 charging stations.

Storage incentives are another policy option that could support investment in EV infrastructure. On-site battery storage systems can be installed to allow solar power to be used for EVs. Tax incentives for batteries that facilitate EV use could encourage more of this activity. The Energy Storage Tax Incentive and Deployment Act of 2019 (S. 1142/H.R. 2096) would provide an investment credit for business or home use of energy storage.

Tax-preferred bond financing options could also be used to support EV infrastructure investment.[24] State and local financing for infrastructure solely or partially dedicated to EV charging projects may use tax-exempt bonds so long as the project is classified as serving a "public purpose" as defined in the federal code (IRC §141).[25] A federal infrastructure bank or "green bank" might also be used to support EV infrastructure investment. One option to capitalize this type of bank would be to issue tax-favored bonds.

EVs and Federal Highway Taxes

Since 1956, federal surface transportation programs have been funded largely by taxes on motor fuels that flow into the highway account of the

driver in the state. The same, however, was not found to be true for Tesla Model S purchases. Others have found that the credits for charging infrastructure are not correlated with greater EV adoption. See Alan Jenn, Katalin Springel, and Anand R. Gopal, "Effectiveness of Electric Vehicle Incentives in the United States," *Energy Policy*, vol. 119 (2018), pp. 349-356. Shanjun Li, Lang Tong, Jianwei Xing, and Yiyi Zhou, "The Market for Electric Vehicles: Indirect Network Effects and Policy Design," *Journal of the Association of Environmental and Resource Economists*, March 2017, pp. 89-133, found that, in some circumstances, it was more cost effective to subsidize EV infrastructure than to provide tax incentives for EV purchases.

[24] For a general overview of infrastructure investment financing options, see CRS Report R43308, *Infrastructure Finance and Debt to Support Surface Transportation Investment*, by William J. Mallett and Grant A. Driessen.

[25] For more information on tax-exempt bonds, see CRS Report RL30638, *Tax-Exempt Bonds: A Description of State and Local Government Debt*, by Grant A. Driessen.

Highway Trust Fund (HTF).[26] A steady increase in the revenues flowing into the HTF due to increased motor vehicle use and occasional increases in fuel tax rates accommodated growth in surface transportation spending over several decades. In 2001, though, trust fund revenues stopped growing faster than spending. In 2008, Congress began providing Treasury general fund transfers to keep the highway account solvent.

Electric vehicles do not burn motor fuels, and hence users do not pay motor fuels taxes. Several states have imposed some form of tax or fee on electric vehicles that is dedicated to transportation, such that EV drivers also contribute to paying for highway infrastructure. Legislation has been introduced in the 116th Congress—the Fairness for Every Driver Act (S. 343/H.R. 1027)—that would, in addition to repealing the existing tax credit for plug-in electric vehicles, impose an annual fee on alternative-technology vehicles that draw power from a source not subject to fuel excise taxes.[27] This fee would be designed to compensate for plug-in electric vehicles not paying the gas tax (or other fuel excise tax). Imposing a fee on electric vehicles or other alternative vehicles would increase their cost of ownership, although as a share of the vehicle's total cost, the amount would likely be small.[28] Exempting electric vehicles from taxes or fees imposed on other types of vehicles is one option for encouraging the purchase of electric vehicles.

---

[26] Fuel taxes provide as much as 90% of the Highway Trust Fund's revenue. For more information, see CRS Report R45350, *Funding and Financing Highways and Public Transportation*, by Robert S. Kirk and William J. Mallett.

[27] Under the Fairness for Every Driver Act, a federal "alternative fuel vehicle highway user fee" would be imposed on each vehicle's owner, who would pay the fee when filing annual federal income taxes.

[28] A fee for electric vehicles could be designed to raise an amount of revenue similar to what each conventional vehicle pays in federal motor fuel excise taxes per year. Light duty vehicles consumed an average of 522 gallons in 2016. If this consumption was taxed at 18.3 cents per gallon, federal revenue raised was $95.50 per vehicle. For fuel consumption statistics, see Federal Highway Administration, *Highway Statistics 2016*, "Annual Vehicle Distance Traveled in Miles and Related Data—2016, by Highway Category and Vehicle Type," revised May 2018, at https://www.fhwa.dot.gov/policyinformation/statistics/2016/vm1.cfm.

RESEARCH AND DEVELOPMENT PRIORITIES

Investment in transportation electrification R&D is one approach to reducing the overall cost of electric vehicle technologies. The Obama Administration made vehicle electrification a national goal—including increased spending on battery R&D, stimulus funding for construction of battery manufacturing plants in the United States, and development of DOE electric vehicle research and demonstration programs such as the "EV Everywhere Grand Challenge."[29] The Trump Administration requested reductions in funding levels for vehicle technologies R&D for FY2018 and FY2019, but Congress instead increased the program's funding levels for those years.

Federal R&D funding for electric vehicles and electric vehicle charging infrastructure is primarily administered through the DOE's Office of Energy Efficiency and Renewable Energy (EERE).[30] Within EERE, the Office of Transportation oversees the sustainable transportation R&D portfolio, which includes R&D programs in vehicle technologies, bioenergy technologies, and hydrogen and fuel cell technologies. Activities related to electric vehicles and charging infrastructure are within the Vehicle Technologies Office.

---

[29] In his 2011 State of the Union address, President Obama called for 1 million plug-in electric vehicles on the road by 2015; that objective was met in 2018. The Obama Administration goals were outlined in U.S. Department of Energy, *One Million Electric Vehicles by 2015: February 2011 Status Report*, p. 2, at https://www1.eere.energy.gov/ vehiclesandfuels/ pdfs/1_million_electric_vehicles_rpt.pdf. The DOE program was announced by President Obama in March 2012 with a goal for the United States to be the first country to build plug-in electric vehicles affordable for the average U.S. family by 2022. U.S. Department of Energy, *EV Everywhere: Grand Challenge Blueprint*, January 31, 2013, at https://www.energy.gov/sites/prod/files/2014/02/f8/eveverywhere_blueprint.pdf.

[30] EERE supports renewable energy and end-use energy efficiency technology research, development, and implementation. EERE is led by the Assistant Secretary for Energy Efficiency and Renewable Energy, and it is organized into four offices: Office of Transportation, Office of Renewable Power, Office of Energy Efficiency, and Office of Operations. DOE was established under the Department of Energy Organization Act of 1977 (P.L. 95-91). Section 203 of the act specifies eight assistant secretary positions, whose functions may be assigned by the Secretary of Energy. DOE, Energy Efficiency and Renewable Energy, "Office of Energy Efficiency and Renewable Energy Organization Chart," January 17, 2019, at https://www.energy.gov/eere/downloads/office-energy-efficiency-andrenewable-energy-organization-chart.

In the 116th Congress, some legislative proposals would support R&D programs. The Vehicle Innovation Act of 2019 (S. 1085/H.R. 2170) would authorize R&D programs and other activities to develop innovative vehicle technologies (including electric vehicles and charging infrastructure).

Appropriations for Electric Vehicle Research

Congress provides funding for electric vehicle technologies R&D through annual appropriations to EERE in the Energy and Water Development appropriations bill. The Trump Administration's budget request for EERE was $343 million for FY2020, $2,036 million (86%) less than the FY2019 enacted level of $2,379 million[31] (*Table 1*). The budget request, which "focuses DOE resources toward early-stage R&D and reflects an increased reliance on the private sector to fund later-stage research, development, and commercialization of energy technologies," would provide $73.4 million to vehicle technologies for FY2020, $279.6 million (79%) less than the $344 million that was directed to vehicle technologies within the Joint Explanatory Statement of the FY2019 appropriations conferees.[32]

The FY2019 joint explanatory statement included support for various vehicle technologies to advance transportation electrification research:

- $163.2 million for the battery and electrification technologies subprogram ($38.1 million for electric drive research and development—including $7 million to enable extreme fast charging and advanced battery analytics);

[31] President Trump submitted his FY2020 detailed budget proposal to Congress on March 18, 2019 (after submitting a general budget overview on March 11), and DOE released a detailed FY2020 budget request in March 2019. The budget request for EERE can be found in DOE, *FY 2020 Congressional Budget Justification*, vol. 3, part 2, March 2019, at https://www.energy.gov/sites/prod/files/2019/04/f61/doe-fy2020-budget-volume-3-Part-2.pdf (hereinafter "DOE, *FY 2020 Budget Justification*, vol. 3, part 2")

[32] DOE, FY 2020 Budget Justification, vol. 3, part 2, p. 9; Joint Explanatory Statement of the Committee of Conference on H.R. 5895, Division A—Energy and Water Development and Related Agencies Appropriations Act, 2019.

- $10 million for continued funding of Section 131 of the 2007 Energy Independence and Security Act (P.L. 110-140) for transportation electrification; and
- $37.8 million for the Clean Cities program, which provides competitive grants to support alternative fuel, infrastructure, and vehicle deployment activities.

Table 1. EERE FY2016-FY2019 Enacted Appropriations and FY2020 Request (in millions of current dollars)

		Enacted			Request
	FY2016	FY2017	FY2018	FY2019	FY2020
EERE, Total	2,069.2	2,090.2	2,321.8	2,379.0	695.6
Sustainable Transportation	636.0	613.0	674.0	690.0	163.5
Vehicle Technologies	310.0	307.0	337.5	344.0	68.5
Bioenergy Technologies	225.0	205.0	221.5	226.0	37.0
Hydrogen and Fuel Cell Technologies	101.0	101.0	115.0	120.0	58.0

Source: P.L. 115-244 Division A explanatory statement; P.L. 115-31 Division D explanatory statement.
Note: EERE = DOE's Office of Energy Efficiency and Renewable Energy.

According to the FY2020 DOE budget request, the vehicle technologies program has set goals that "are necessary for new technology options to be more efficient and at least as affordable compared to [the] baseline while also accounting for consumer pay-back period expectations."[33] To advance vehicle electrification, DOE established the following research priorities:

- identify new battery chemistry and cell technologies with the potential to reduce the cost of electric vehicle batteries by more than half, to less than $100 per kilowatt hour (kWh) (with an ultimate goal of $80/kWh);

[33] DOE considers the "baseline" lifecycle cost of a future (circa 2030) modeled conventional gasoline internal combustion engine vehicle to be approximately 27 cents per mile and 4,700 Btu per mile. This baseline establishes a common metric for research programs within Sustainable Transportation for comparison; see DOE, *FY2020 Budget Justification*, vol. 3, part 2, p. 17.

- increase vehicle range to 300 miles; and
- decrease charge time to 15 minutes or less by 2028.[34]

The request for FY2020 would reduce funding and prioritize "early-stage activities," including the development of critical materials-free battery technologies. The request would eliminate funding for battery safety testing, battery thermal performance testing, and Clean Cities coalition support, training and technical assistance, and partnership activities.[35]

Clean Cities Program

The DOE Clean Cities program supports local actions to reduce petroleum use in transportation.[36] The program funds transportation projects nationwide through a competitive application process, and leverages these funds with additional public- and private-sector matching funds and in-kind contributions.[37] While the program supports a variety of alternative fuels and vehicles in an effort to reduce petroleum use, funding opportunities by Clean Cities that directly support EVs include the EV Everywhere Plug-In Electric Vehicle Local Showcases.[38]

---

[34] DOE, *FY2020 Budget Justification*, vol. 3, part 2, p. 18. For reference, a survey found that commercially available lithium-ion battery packs in 2018 ranged in price between $126/kWh to $460/kWh with a volume-weighted average of $176/kWh; see James Frith, *2018 Lithium-Ion Battery Price Survey*, BloombergNEF, December 19, 2018, p. 2.

[35] DOE, *FY2020 Budget Justification*, vol. 3, part 2, pp. 20, 46.

[36] Office of Energy Efficiency and Renewable Energy, *Clean Cities: Building Partnerships to Cut Petroleum Use in Transportation*, DOE/GO-102015-4743, November 2015, at https://www.afdc.energy.gov/uploads/publication/clean_cities_overview.pdf.

[37] Since 1993, the program has funded nearly 600 transportation projects and distributed approximately $400 million in project awards. These funds have leveraged nearly $800 million in additional public- and private-sector matching funds and in-kind contributions; see DOE, Energy Efficiency and Renewable Energy, Clean Cities Coalition Network, "Goals and Accomplishments," at https://cleancities.energy.gov/accomplishments/.

[38] U.S. Department of Energy, Office of Energy Efficiency and Renewable Energy, Clean Cities Coalition Network, "Funded Projects," at https://cleancities.energy.gov/partnerships/projects. In addition to the EV Everywhere Plug-In Electric Vehicle Local Showcase, the Electric Vehicle Community Readiness program provided one-time funding of $8.5 million in 2011 to 16 projects to help communities prepare for plug-in electric vehicles and charging infrastructure. As part of the readiness program, community partnerships collaborated on plans to deploy EVs, including activities such as streamlining permitting processes, revising building codes, training emergency personnel, educating the public, and developing incentives. U.S. Department of Energy, Office of Energy Efficiency and

EV Everywhere Plug-In Electric Vehicle Local Showcases were selected in 2016 and are in progress. The three projects were selected to promote and demonstrate plug-in electric vehicle (PEV) use by "establishing local showcases that provide a hands-on consumer experience and in-depth education in a conveniently located, brand-neutral setting."[39] The awardees plan to establish showcases in at least 14 states, including states in the Upper Midwest, California, the Pacific Northwest, and New England.

In the FY2019 joint explanatory statement, Congress directed DOE "to continue to support the Clean Cities program, including competitive grants to support alternative fuel, infrastructure, and vehicle deployment activities."[40] In the FY2020 budget request, the Trump Administration proposed eliminating funding for the Clean Cities program.[41] In the 116th Congress, two bills that pertain to electric vehicles would establish competitive grant programs within the Department of Transportation for electric vehicle charging infrastructure (H.R. 2616 and S. 674).

Alternative Fuel Corridors

Purchases of electric vehicles in the near future will depend to some extent on the steps state and local governments and private entities take to build a reliable network of charging stations. Section 1413 of the Fixing America's Surface Transportation Act (FAST Act; P.L. 114-94) seeks to address that goal; it requires the Department of Transportation (DOT) to designate by 2020 national alternative fuel corridors (AFCs) to promote vehicle use of electricity, hydrogen, propane, and natural gas. The Federal Highway Administration (FHWA) has been working with other federal,

---

Renewable Energy, Alternative Fuels Data Center, "Plug-In Vehicle Readiness," at https://afdc.energy.gov/pevreadiness.html.

39 U.S. Department of Energy, Office of Energy Efficiency and Renewable Energy, Clean Cities Coalition Network, "Funded Projects," at https://cleancities.energy.gov/partnerships/projects.

40 Joint Explanatory Statement of the Committee of Conference on H.R. 5895, Division A—Energy and Water Development and Related Agencies Appropriations Act, 2019.

41 DOE, *FY2020 Budget Justification*, vol. 3, part 2, p. 46.

state, and local officials and industry groups to plan AFC designations on interstate corridors (*Figure 2*). FHWA has assigned designations to highways as either "corridor ready"— they have enough fueling stations to serve a corridor—or "corridor pending," where alternative fueling is insufficient. In the case of electric vehicles, a corridor-ready designation would apply if there were EV charging stations at 50-mile intervals, with a goal of establishing Level 3 DC Fast Charge infrastructure.

Under this program, FHWA has developed standardized AFC signage and other forms of public education, and encouraged regional cooperation in planning new fueling networks. FHWA has undertaken three rounds of AFC nominations, the latest announced in April 2019. FHWA has identified building out alternative corridors on the most traveled Interstates, such as I-95 and I-80, as priorities for third-round funding. An additional goal is to secure nominations for areas targeted for EV investments by Electrify America in its Zero Emission Vehicle (ZEV) investment plan.[42] Signage and installation of alternative fueling infrastructure along these corridors have been determined to be eligible expenses under FHWA's Congestion Mitigation and Air Quality Improvement (CMAQ) Program.[43] In addition, the Department of Energy's Clean Cities program provides funding for fueling infrastructure, and some states have similar programs.

Federal Support for Municipal Bus Electrification

Until recently, the operation of battery electric buses in U.S. cities was seen as a long-term prospect because of their relatively high cost, range

---

[42] Electrify America was formed in 2016 as part of Volkswagen Group's settlement with the U.S. Department of Justice. It is a subsidiary of Volkswagen Group.

[43] Program descriptions provided to CRS in FHWA email of November 5, 2018; also, see Federal Highway Administration, "Interim Program Guidance Under MAP-21," November 12, 2013, at https://www.fhwa.dot.gov/environment/air_quality/cmaq/policy_and_guidance/2013_guidance/index.cfm. In addition, the Manual on Uniform Traffic Control Devices (MUTCD) provides guidance on signage design and placement. See Federal Highway Administration, "Signing for Designated Alternative Fuels Corridors," in *Manual on Uniform Traffic Control Devices (MUTCD)*, at https://mutcd.fhwa.dot.gov/resources/policy/alt_fuel_corridors/index.htm.

limits, and recharging infrastructure needs. But with technological improvements, public transportation agencies have begun to show interest in electric buses to replace vehicles powered by diesel and other fossil fuels. This interest is especially strong in metropolitan areas with air quality problems. The Federal Transit Administration (FTA) provides substantial support to transit agencies to purchase buses. Federal funds can be spent on most types of bus technology; the choice of technology is up to the transit agency concerned.

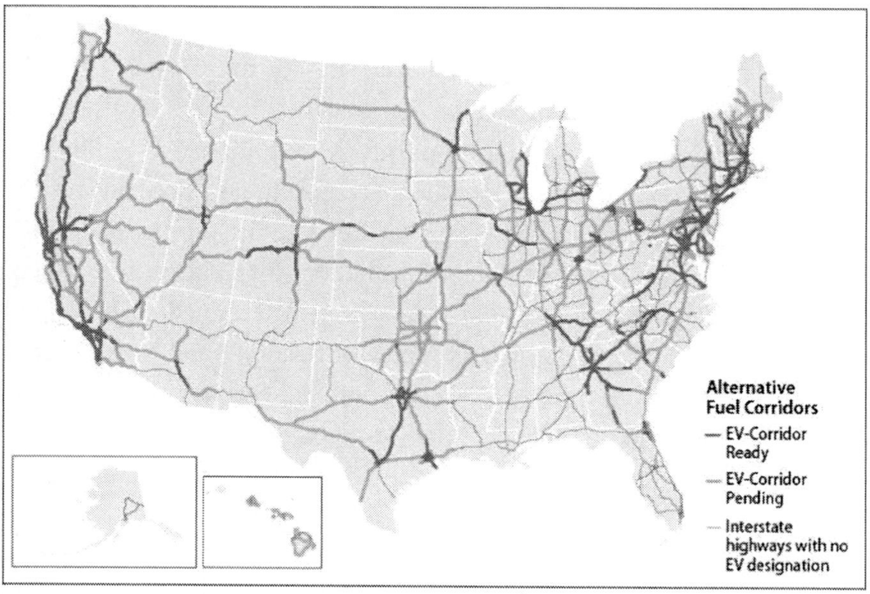

Source: CRS map based on Federal Highway Administration data at https://hepgis.fhwa.dot.gov/fhwagis/ViewMap.aspx?map=Highway+Information|Electric+Vehicle+(EV-Round+1,+2+and+3)#.

Notes: States can designate portions of highways in their jurisdiction as alternative fuel corridors, subject to approval by FHWA, which approved Rounds 1, 2, and 3 of such designations in 2016, 2017, and 2018. These rounds have resulted in corridors that are designated as "ready" or "pending," categories based on the maximum distance between fueling stations. For example, if EV charging stations are more than 50 miles apart, that portion of highway is designated "pending."

Figure 2. Alternative Fuel Corridors for Electric Vehicles. (Corridors in 2019)

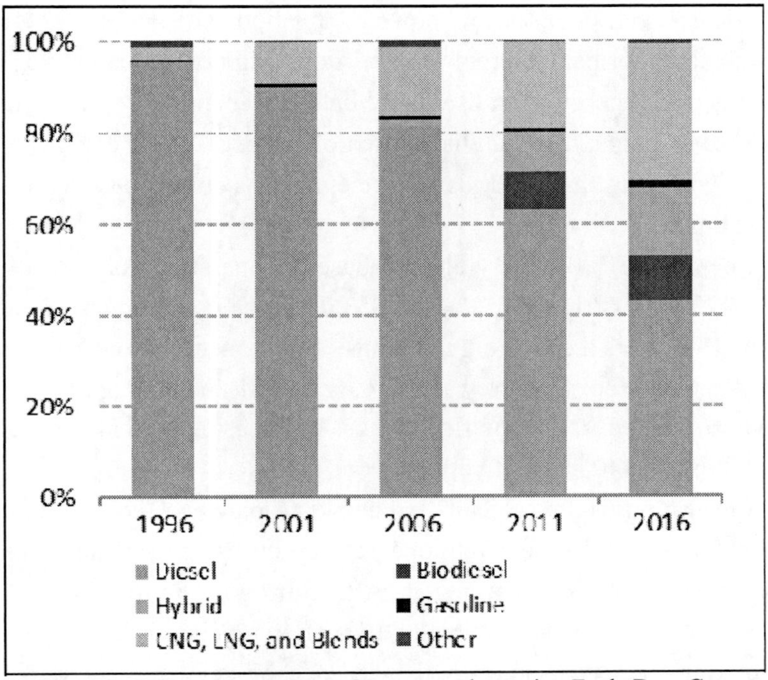

Source: CRS, based on U.S. Department of Energy, Alternative Fuels Data Center, and American Public Transportation Association (APTA) Fact Book and Appendix, 2017.

Notes: Data show buses in operation. "Other" includes buses propelled with electricity, hydrogen, and propane.

Figure 3. U.S. Transit Buses by Fuel Type.

Transit buses typically operate over short distances with fixed routes and frequent stops. In 1996, 95% of the buses in service were powered by diesel fuel (Figure 3). More recently, transit agencies have integrated buses fueled by compressed natural gas (CNG), liquefied natural gas (LNG), and biodiesel into their fleets. Since the end of the last recession, the share of lower-emission hybrid buses—including diesel buses with electric motors—has also increased, rising from just under 5% of buses in use in 2009 to nearly 15% in 2016.

Diesel-electric buses are powered by both an electric motor and a smaller-than-normal internal combustion engine; regenerative braking systems store energy from use of the bus's mechanical systems, giving the bus greater range. The purchase price of hybrids is less expensive than a fully electric bus, and hybrids reduce emissions compared to conventional diesel buses.[44]

There were 300 battery electric buses in operation domestically at the end of 2017, less than 0.5% of the 65,000 buses in public transit agencies' fleets. However, the two biggest transit bus systems in the United States, Los Angeles Metro and New York City Transit, have announced plans to move to zero-emission bus fleets, most likely using battery electric buses, by 2030 and 2040, respectively.[45]

Electric buses are typically expensive to purchase, costing as much as $300,000 more than conventional diesel buses, and require additional investment to build recharging infrastructure. On the other hand, electric buses are quieter than internal combustion engine vehicles, may have lower operating costs due to the absence of engines and transmissions requiring maintenance, and have low or zero direct emissions. The range an electric bus can travel on one charge has in the past been a limiting factor, but newer models can travel more than 200 miles, still short of the 600-mile or more range that conventional and other alternative fuel buses can travel.[46] A study by Carnegie Mellon University found that when social costs, such as the health effects of diesel emissions, are taken into

---

[44] Carnegie Mellon University, Scott Institute for Energy Innovation, *Which Alternative Fuel Technology Is Best for Transit Buses?*, January 2017, p. 19, at https://www.cmu.edu/energy/education-outreach/public-outreach/17-104%20Policy%20Brief%20Buses_WEB.pdf.

[45] Los Angeles County Metropolitan Transportation Authority, "Metro Leads the Nation in Setting Ambitious 2030 Zero Emission Bus Goal," press release, August 2, 2017, at https://www.metro.net/news/simple_pr/metro-leads-setting2030-zero-emission-bus-goal/; and New York Metropolitan Transportation Authority (MTA), *MTA Board Meeting*, April 25, 2018, remarks by Andy Byford, president of New York City Transit Authority, at 1:17:14 in the following video of the MTA board meeting: https://www.youtube.com/watch?v=3u2Z35Awh84.

[46] According to the Carnegie Mellon University study, 40-foot buses have typical ranges of 600 miles for compressed natural gas, 690 miles for diesel, and 720 for hybrid-electric. Carnegie Mellon University Scott Institute for Energy Innovation, *Policymaker Guide: Which Alternative Fuel Technology is Best for Transit Buses?*, January 2017, p. 15.

account, battery electric buses have lower total annualized costs than conventional diesel buses over the typical 12-year life cycle of a transit bus.[47]

Electric buses are generally eligible for FTA funding under several programs, including the Bus and Bus Facilities Program.[48] One discretionary component of the FTA bus program is the Low or No Emission Vehicle (Low-No) program, which provides funding to state and local authorities for the purchase or lease of zero-emission and low-emission transit buses as well as acquisition, construction, and leasing of required supporting facilities. Electric buses are purchased through the Low-No bus program; mandatory spending of $55 million per year through FY2020 was authorized in the FAST Act. An additional $29.5 million was appropriated for FY2018.

School buses generally have diesel engines, and they are primarily funded locally. FTA does not fund school buses, but the Environmental Protection Agency (EPA) administers the School Bus Rebate Program, which assists school districts in reducing diesel emissions. In 2017, nearly $9 million was provided to replace diesel buses with diesel-electric hybrids.[49]

In the 116th Congress, some legislative proposals would support vehicle fleet electrification, including transit buses. The Green Bus Act of 2019 (H.R. 2164) would require that any bus purchased or leased with funds provided by the Federal Transit Administration for public transportation purposes be a zero-emission bus. The Federal Leadership in Energy Efficient Transportation (FLEET) Act (H.R. 2337) would authorize the U.S. Postal Service to enter into energy savings performance contracts to purchase or lease low-emission and fuel-efficient vehicles (including

---

[47] In the Carnegie Mellon University study, total annualized costs per bus are determined by summing life cycle agency costs (such as bus purchase costs, fuel costs, and operation and maintenance costs) with social costs (such as costs from climate change and air pollution). Carnegie Mellon University, Scott Institute for Energy Innovation, *Which Alternative Fuel Technology Is Best for Transit Buses?*, January 2017, p. 18.

[48] The FTA Bus and Bus Facilities Program primarily distributes funds to local transit systems by formula.

[49] U.S. Environmental Protection Agency, "EPA School Bus Rebate Program to Reduce Diesel Emissions," press release, October 16, 2018, at https://www.epa.gov/newsreleases/epa-school-bus-rebate-program-reduce-dieselemissions.

electric vehicles) and to construct or maintain infrastructure, including electric vehicle charging stations, among other provisions.[50]

OTHER FEDERAL POLICIES AFFECTING ELECTRIFICATION

The following policies were generally established for purposes not directly related to vehicle electrification, but they have dimensions that may support that goal.

Volkswagen Settlement

In 2016, Volkswagen Group reached a number of legal settlements concerning violations of the Clean Air Act.[51] As part of its settlement terms, Volkswagen pledged to invest $2 billion over a 10-year period in zero-emissions vehicle infrastructure and education in select U.S. cities through its Electrify America initiative. Of that amount, $800 million is to be spent in California. As an additional condition, Volkswagen was also required to fund a $2.7 billion national Environmental Mitigation Trust, funds from which are available to states and other beneficiaries for mitigating the negative impacts of the excess diesel emissions that were released by Volkswagen's noncompliant vehicles. States could choose to spend some of this special funding on bus electrification, including school buses.

---

[50] For more information on energy savings performance contracts, see CRS Report R45411, *Energy Savings Performance Contracts (ESPCs) and Utility Energy Service Contracts (UESCs)*, by Corrie E. Clark.

[51] Environmental Protection Agency, "EPA Approved National ZEV Investment Plan," press release, April 9, 2017, at https://www.epa.gov/enforcement/epa-approved-national-zev-investment-plan-public-version; and CRS Report R44372, *Volkswagen, Defeat Devices, and the Clean Air Act: Frequently Asked Questions*, by Bill Canis et al.

Federal Motor Vehicle Environmental Regulations

The first motor vehicle fuel economy standards were enacted in 1975 in response to the oil embargo of 1973-1974 (Energy Policy and Conservation Act of 1975; P.L. 94-163, as amended). The National Highway Traffic Safety Administration (NHTSA) sets and enforces the Corporate Average Fuel Economy (CAFE) standards for passenger cars and light trucks, which were most recently legislatively set in the Energy Independence and Security Act of 2007 (P.L. 110-140) at 35 miles per gallon (mpg) by 2020. Because EPA has authority to regulate greenhouse gas (GHG) emissions, it joined with NHTSA during the Obama Administration to promulgate new, stricter combined CAFE/GHG standards for motor vehicles. These new standards established targets for reduced GHG emissions and rising CAFE fuel economy of nearly 50 mpg by model year (MY) 2025. Electrification was seen as one means of reaching the new CAFE/GHG targets.[52]

The Trump Administration has proposed to leave the current CAFE/GHG standards in place through model year 2026, based on its analysis that economic and technological factors have changed since the standards were put in place in 2010.[53] EPA asserted in April 2018 that the goals established for MY2026 could be achieved only with more extensive vehicle electrification than now anticipated, and rejected previous EPA determinations that EV sales in future years would grow rapidly enough to enable automakers to comply with the MY2022-MY2025 standards: "Based on consideration of the information provided, the Administrator believes that it would not be practicable to meet the MY 2022–2025

---

[52] For background and analysis of these standards and Trump Administration recommendations for changes, see CRS Report R45204, *Vehicle Fuel Economy and Greenhouse Gas Standards: Frequently Asked Questions*, by Richard K. Lattanzio, Linda Tsang, and Bill Canis.

[53] NHTSA and EPA, "The Safer Affordable Fuel Efficient (SAFE) Vehicles Proposed Rule for Model Years 2012- 2016," 83 *Federal Register* 42986, August 24, 2018, at https://www.govinfo.gov/content/pkg/FR-2018-08-24/pdf/2018-16820.pdf. The notice of proposed rulemaking asked for comments by October 23, 2018.

emission standards without significant electrification and other advanced vehicle technologies that lack a requisite level of consumer acceptance."[54]

PROSPECTS FOR ELECTRIFICATION OF AUTONOMOUS VEHICLES

Fully autonomous passenger vehicles hold the potential for safer transportation and new mobility for the elderly, the disabled, and those who cannot afford to purchase a car. The projected timeline for emergence of fully autonomous vehicles varies from several years to several decades.

It is often assumed that autonomous vehicles—including those providing ride-sharing services— will employ electric motors instead of internal combustion engines, but that assumption is based on the types of federal and state policies that are put in place. In the absence of policies favoring EVs, future autonomous vehicles could be powered by fossil fuels.

There are several reasons why electrification may not be the ultimate choice for autonomous vehicles. Analyses by Ford and Volvo have for the following reasons led them to retain internal combustion engines for their ride-sharing ventures for the present:[55]

- Expense and Inconvenience. Electric vehicles remain expensive compared to conventional vehicles, and many models have short driving ranges and long refueling times. Shared vehicles, which may be on the road for many hours a day, magnify these shortcomings.
- Battery Life. Frequent use of battery fast charging is known to lead to faster degradation of the battery.

[54] EPA, "Mid-Term Evaluation of Greenhouse Gas Emissions Standards for Model Year 2022–2025 Light-Duty Vehicles," 83 *Federal Register* 16077, April 13, 2018.

[55] Peter Slowik, *The Future of Transportation: Autonomous and ... Internal Combustion?*, International Council on Clean Transportation (ICCT), January 16, 2018, at https://www.theicct.org/blog/staff/future-transportationautonomous-internal-combustion.

- Energy-Intensity. Autonomous vehicles rely on energy-intensive technologies. It has been estimated that operation of an autonomous driving system for two hours in an electric vehicle could use as much as 10% of its stored energy before the vehicle moves, requiring even more frequent recharging.

These drawbacks are countered by the following factors that could make electrification of autonomous vehicles the more attractive power source:

- Compatibility. The extensive use of computers and sensors in autonomous vehicles may be easier to embed in electric vehicles, which have fewer mechanical—and more electronic—parts than a conventional vehicle.
- Operability. Electricity is generally less expensive than gasoline or diesel, and maintenance costs for electric vehicles, with many fewer parts, may be considerably lower.
- Emissions. The power demands of autonomous vehicle equipment and computers will reduce the fuel economy of gasoline and diesel vehicles and increase emissions. It has been estimated that increased power requirements will increase emissions from combustion vehicles by over 60 grams of CO_2 equivalent per mile, equal to reducing fuel economy of a 35 mpg vehicle to 29 mpg. Switching to electricity would diminish the direct emissions.[56]

STATE INCENTIVES AND UTILITY ISSUES

Acceptance of electric vehicles and related infrastructure is affected by legislation, regulations, and policies adopted by state agencies and electric utilities.

---

[56] David Reichmuth, *How Important Is it for Self-Driving Cars to be Electric?*, Union of Concerned Scientists, May 3, 2018, at https://blog.ucsusa.org/dave-reichmuth/how-important-is-it-for-self-driving-cars-to-be-electric.

State Incentives

Incentives vary widely from state to state. The National Conference of State Legislatures tracks vehicle and charger incentives on a state-by-state basis.[57] Forty-five states and the District of Columbia currently offer incentives for certain hybrid or electric vehicles, or both. Those incentives include

- permitting solo drivers of electric and hybrid vehicles to use high-occupancy (carpool) lanes,
- income tax credits and rebates for the purchase of an electric vehicle,
- reduced registration fees,
- parking fee exemptions,
- excise tax and emission test waivers, and
- income tax credits for installation of a home or business charger.[58]

These incentives have been found to vary in their effectiveness. Several analyses have shown that tax incentives for electric vehicles and infrastructure are the "dominant factors in driving PEV adoption."[59] Rebates—which happen at the point of sale or within a short time after a vehicle purchase—have been identified as the most effective incentive because their value is clear to buyers at the time of a vehicle transaction.[60]

---

[57] For a description of incentives in each state, see the National Conference on State Legislatures, "State Efforts to Promote Hybrid and Electric Vehicles," September 26, 2017, at http://www.ncsl.org/research/energy/state-electricvehicle-incentives-state-chart.aspx.

[58] For example, the Massachusetts Electric Vehicle Incentive Program provides grants of 50% of the cost, up to $25,000, to workplaces that install a charging station.

[59] Easwaran Narassimhan and Caley Johnson, "The Role of Demand-Side Incentives and Charging Infrastructure on Plug-In Electric Vehicle Adoption: Analysis of US States," IOPScience, Environmental Research Letters, vol. 13, no. 7, July 13, 2018, p. 9, at https://iopscience.iop.org/article/10.1088/1748-9326/aad0f8.

[60] Ibid. See also Zifei Yang, Peter Slowick, and Nic Lutsey, et al., *Principles for Effective Electric Vehicle Incentive Design*, International Council on Clean Transportation (ICCT), June 2016, at https://www.theicct.org/publications/principles-effective-electric-vehicle-incentive-design.

Zero Emission Vehicle Program

The California Air Resources Board (CARB) adopted low-emission vehicle regulations in 1990, requiring automakers to sell light vehicles in that state that meet progressively cleaner emissions standards.[61] As part of these emission regulations, CARB also established the Zero-Emission Vehicle (ZEV) program, which requires automakers to offer for sale the lowest-emission vehicles available, with a focus on battery electric, plug-in hybrid electric, and hydrogen fuel cell vehicles. The number of ZEVs each automaker is required to sell is based upon its total light-vehicle sales in California. CARB has set ZEV sales percentages through a vehicle credit system, increasing annually to 2025. Nine other states have adopted the California ZEV regulations. The states affected by the regulations represent over one-third of all U.S. new light vehicle sales.[62]

Utilities and EVs and EV Infrastructure: Tax and Regulatory Issues

Electric utilities, which are regulated by the state in which they are located, are in a unique position as the primary providers of electricity to aid in integrating EVs into the grid. Utilities can provide incentives to consumers to charge EVs during off-peak hours when there is excess generation. Utilities may also be in the position to install public electric charging infrastructure, assuming that there are no limits on their owning these assets. Many of the barriers utilities face with respect to electrification infrastructure are regulatory, and the role of tax policy in addressing such barriers may be limited.

Sluggish growth in energy demand has posed a challenge for the electric utility sector. The industry has recognized EVs as an opportunity

---

[61] California Air Resources Board, *Zero-Emission Vehicle Program*, viewed May 13, 2019, at https://ww2.arb.ca.gov/our-work/programs/zero-emission-vehicle-program/about.

[62] The Clean Air Act enables other states to adopt California's standards. The nine states are Connecticut, Maine, Maryland, Massachusetts, New York, New Jersey, Oregon, Rhode Island, and Vermont.

for growth. For this reason, electric utilities may have their own market-based incentives (absent federal intervention) to invest in EV infrastructure and take measures to support consumer EV adoption. Further, electric utilities may support extending current-law vehicle and infrastructure tax credits.[63] At the same time, preparing the grid for a surge in electric car ownership could require substantial capital investments, if peak demand is increased.

Tax Incentives to Utilities for Electric Vehicle Infrastructure

A number of challenges are associated with providing tax incentives to utilities that provide EV infrastructure. While 65% of electricity customers across the United States are customers of investor-owned utilities, most other users purchase electricity from cooperatives or municipal power providers. There are limited options for providing a direct federal tax benefit to cooperative or municipal utilities that do not pay federal income taxes.[64] Further, in many states customers may purchase electricity from competing suppliers and pay the local electric utility for delivering it.[65] A tax incentive to provide EV infrastructure might be made available only to utilities, to other electricity suppliers, or to both. Applicability of such an incentive would vary according to state policies or the type of utility.[66]

State regulatory commissions typically establish prices or rates that allow utilities to earn a rate of return that the regulator determines to be reasonable. This rate of return is fixed, such that additional tax incentives do not necessarily increase the return that a utility can realize. Federal taxes are an operating expense. In some states, tax incentives that reduce utilities' operating expenses result in lower rates for electricity customers

---

[63] Samantha Raphelson, "U.S. Utilities Look to Electric Cars as Their Savior Amid Decline in Demand," NPR, March 2018, at https://www.npr.org/2018/03/29/598032288/u-s-utilities-look-to-electric-cars-as-their-savior-amid-decline-indemand.

[64] Data on utility bundled sales by ownership type can be found in Table 10 U.S. Energy Information Administration, "Electric Sales, Revenue, and Average Price," at https://www.eia.gov/electricity/sales_revenue_price/.

[65] U.S. Energy Information Administration, "Power Marketers Are Increasing Their Share of U.S. Retail Electricity Sales," June 12, 2018, at https://www.eia.gov/todayinenergy/detail.php?id=36415.

[66] For example, some publicly owned utilities may not be able to take advantage of federal tax incentives.

(not higher returns for utilities). Hence, federal tax incentives may provide a limited near-term incentive for utilities to increase capital spending on EV infrastructure.

State Regulatory Considerations

In the United States, the sale of electricity is governed by many different federal, state, and local regulations. When it comes to the sale of electricity for the purpose of charging EVs, the states generally have regulatory jurisdiction over retail electricity transactions,[67] though federal and municipal authorities may also play a role.[68] State approaches to regulation vary considerably.

Rules and regulations governing the retail sale of electricity generally originate with a state public utility commission. An electric utility is defined in federal law as any person, state, or federal agency "which sells electric energy."[69] This definition could potentially be interpreted to mean that electric vehicle charging station operators are electric utilities by virtue of the fact that they sell electricity, and are therefore subject to all laws, requirements, and regulations pertaining to electric utilities.

Should charging station operators be subject to regulation as electric utilities, or is regulatory reform necessary to accommodate this new class

---

[67] Jurisdiction over the sale of electricity to or from an EV charging station hinges upon its definition as either a retail transaction or a sale for resale. Retail transactions are generally defined by the Federal Energy Regulatory Commission (FERC) as "sales made directly to the customer that consumes the energy product," whereas sales for resale are defined as "a type of wholesale sales covering energy supplied to other electric utilities, cooperatives, municipalities, and Federal and state electric agencies for resale to ultimate consumers." (FERC, "Glossary," accessed November 6, 2018, at https://www.ferc.gov/resources/glossary.asp?csrt=17301308159773522879#R.) States typically regulate retail electricity transactions, while FERC has jurisdiction over the transmission and wholesale sales of electricity in interstate commerce. The question of how to define sales of electricity to and from charging stations (including vehicleto-grid transactions) may be subject to significant legal interpretation, and potentially represents the intersection of various federal and state statutes and regulations. A comprehensive legal analysis is beyond the scope of this chapter.

[68] To date, FERC has not intervened in state decisions regarding the classification and regulation of EV infrastructure. In the absence of clear or comprehensive state policies, some cities and counties have become the de facto regulators of EV charging stations within their regions.

[69] See §3 of the Public Utility Regulatory Policies Act of 1978 (P.L. 95-617, 16 U.S.C. §2602(4)).

of electricity transactions? Faced with the question of whether or how to regulate the operators, states have taken a variety of approaches:

- Some states have issued new guidelines or regulations that define the requirements for regulated utilities to operate charging stations.[70] For example, some states (e.g., Oregon) allow existing regulated utilities to invest in and operate EV charging stations as separate, nonregulated ventures. Others (e.g., Texas) have effectively limited the operation of charging stations to electric distribution utilities by requiring operators to meet high technical and financial standards. Still others (e.g., Kansas) have prevented electric utilities from owning and operating charging stations altogether.
- Other states (e.g., New York) have exempted charging station operators from public utility regulations. This leaves questions as to which regulatory agency, if any, is responsible for regulating the charging station operators, as well as whether additional regulation is needed in order to ensure fair market practices.[71]
- Finally, some states have refrained from taking action altogether. Without regulatory changes, private charging station operators in these states may be subject to regulation as a utility by the state's public commission. Lack of clarity about how operators will be regulated is seen by some as an impediment to the spread of the technology in these states.[72] In some cases, EV service providers have avoided regulation as a utility by providing charging services for free, or by charging customers by the minute rather than by the amount of electricity used.

---

[70] Martha Moore, "Should Utilities Build Charging Stations for Electric Cars?," Pew, Stateline (blog), September 11, 2017, at http://pew.org/2wOozZk.

[71] For a more detailed discussion of these issues, see Christopher Young, Marc Machlin, and Erica Hall Dressler, "Electric Vehicle Charging Station Regulation: Why Your Local Electric Vehicle Charging Station Doesn't (and Shouldn't) Look Like Your Local Gas Station," Client Alert (Pepper Hamilton LLP), June 23, 2016.

[72] The Council of State Governments, "State Utilities Law and Electric Vehicle Charging Stations," October 2013, at http://knowledgecenter.csg.org/kc/content/state-utilities-law-and-electric-vehicle-charging-stations.

Whether public utilities or private companies may operate electric vehicle charging stations and whether station operators are subject to regulation as a utility may affect deployment of EV charging infrastructure. State jurisdiction over retail electricity transactions may limit the potential role of the federal government in regulating the provision of EV charging services.

An additional consideration is the potential for electric vehicle batteries to be used for storage, referred to as vehicle-to-grid (V2G) storage. V2G storage would allow idle vehicle batteries to be used for grid services, such as demand response. The batteries could reduce vehicle owners' electricity demand during peak periods or provide electricity to the grid during peak periods in response to time-based rates or other financial incentives. In the United States, utilities are beginning to test V2G performance in demonstration projects.[73] In addition to technology, other identified challenges include regulation, market, and end-user acceptance.[74]

---

[73] For example, Austin Energy, a publicly owned utility, through the Austin Sustainable and Holistic Integration of Energy Storage and Solar Photovoltaics (SHINES) project is reportedly testing V2G performance. Justin Gerdes, "Vehicle-to-Grid Testing Comes to Texas," *Greentech Media*, February 14, 2019, at https://www.greentechmedia. Com /articles/read/vehicle-to-grid-testing-austin-energy-pecan-street#gs.c658jp.

[74] Johannes Kester, Lance Noel, and Gerardo Zarazua de Rubens, et al., "Promoting Vehicle to Grid (V2G) in the Nordic Region: Expert Advice on Policy Mechanisms for Accelerated Diffusion," *Energy Policy*, vol. 116 (May 2018), pp. 422-432.

In: Key Congressional Reports for June 2019 ISBN: 978-1-53616-662-0
Editor: Stephanie A. Betts

Chapter 3

CENTRAL VALLEY PROJECT: ISSUES AND LEGISLATION*

Charles V. Stern and Pervaze A. Sheikh

SUMMARY

The Central Valley Project (CVP), a federal water project owned and operated by the U.S. Bureau of Reclamation (Reclamation), is one of the world's largest water supply projects. The CVP covers approximately 400 miles in California, from Redding to Bakersfield, and draws from two large river basins: the Sacramento and the San Joaquin. It is composed of 20 dams and reservoirs and numerous pieces of water storage and conveyance infrastructure. In an average year, the CVP delivers more than 7 million acre-feet of water to support irrigated agriculture, municipalities, and fish and wildlife needs, among other purposes. About 75% of CVP water is used for agricultural irrigation, including 7 of California's top 10 agricultural counties. The CVP is operated jointly with the State Water Project (SWP), which provides much of its water to municipal users in Southern California.

---

* This is an edited, reformatted and augmented version of Congressional Research Service, Publication No. R45342, dated June 21, 2019.

CVP water is delivered to users that have contracts with Reclamation. These contractors receive varying levels of priority for water deliveries based on several factors, including hydrology, water rights, prior agreements with Reclamation, and regulatory requirements. The Sacramento and San Joaquin Rivers' confluence with the San Francisco Bay (*Bay-Delta* or *Delta*) is a hub for CVP water deliveries; many CVP contractors south of the Delta receive water that is "exported" from north of the Delta.

Development of the CVP resulted in significant changes to the area's natural hydrology. However, construction of most CVP facilities predated major federal natural resources and environmental protection laws. Much of the current debate related to the CVP revolves around how to deal with changes to the hydrologic system that were not significantly mitigated for when the project was constructed. Thus, multiple ongoing efforts to protect species and restore habitat have been authorized and are incorporated into project operations.

Congress has engaged in CVP issues through oversight and at times legislation, including provisions in the 2016 Water Infrastructure Improvements for the Nation (WIIN Act; P.L. 114-322) that, among other things, authorized changes to operations in an attempt to provide for delivery of more water under certain circumstances. Although some stakeholders are interested in further operational changes to enhance CVP water deliveries, others are focused on the environmental impacts of operations.

Various state and federal proposals are currently under consideration and have generated controversy for their potential to affect CVP operations and allocations. In late 2018, the State of California finalized revisions to its Bay-Delta Water Quality Control Plan. These changes would require that more flows from the San Joaquin and Sacramento Rivers reach the Bay-Delta for water quality and fish and wildlife enhancement (and thus would further restrict water supplies for other users). The changes have generally been opposed by the Trump Administration. At the same time, the Trump Administration is pursuing efforts to increase CVP water supplies for users, including changes to CVP operations under an October 2018 White House memorandum on western water supplies. Efforts to add or supplement CVP storage and conveyance are also being considered: The state is proposing a new water conveyance project (known as the California WaterFix) that would bypass the Bay-Delta and, under certain conditions, increase exports from north to south for some users. Additionally, new storage projects are under study by federal and state entities; these projects would aim to increase CVP and/or SWP water supplies.

In the 116th Congress, legislators may consider bills and conduct oversight on efforts to increase CVP water exports compared to current baselines. Congress is considering whether to approve funding for new water storage projects, and may also consider legislation to extend or amend previously enacted CVP authorities (e.g., WIIN Act authorities that are expiring or have exceeded their appropriations ceiling).

INTRODUCTION

The Bureau of Reclamation (Reclamation), part of the Department of the Interior (DOI), operates the multipurpose federal Central Valley Project (CVP) in California, one of the world's largest water storage and conveyance systems. The CVP runs approximately 400 miles in California, from Redding to Bakersfield (Figure 1). It supplies water to hundreds of thousands of acres of irrigated agriculture throughout the state, including some of the most valuable cropland in the country. It also provides water to selected state and federal wildlife refuges, as well as to some municipal and industrial (M&I) water users.

This chapter provides information on hydrologic conditions in California and their impact on state and federal water management, with a focus on deliveries related to the federal CVP. It also summarizes selected issues for Congress related to the CVP.

RECENT DEVELOPMENTS

The drought of 2012-2016, widely considered to be among California's most severe droughts in recent history, resulted in major reductions to CVP contractor allocations and economic and environmental impacts throughout the state.[1] These impacts were of interest to Congress, which oversees federal operation of the CVP. Although the drought ended with the wet winter of 2017, many of the water supply controversies associated with the CVP predated those water shortages and remain unresolved. Absent major changes to existing hydrologic, legislative, and

---

[1] For more information on drought in general, see CRS Report R43407, *Drought in the United States: Causes and Current Understanding*, by Peter Folger.

regulatory baselines, most agree that at least some water users are likely to face ongoing constraints to their water supplies. Due to the limited water supplies available, proposed changes to the current operations and allocation system are controversial.

As a result of the scarcity of water in the West and the importance of federal water infrastructure to the region, western water issues are regularly of interest to many lawmakers. Legislation enacted in the 114th Congress (Title II of the Water Infrastructure Improvements for the Nation [WIIN] Act; P.L. 114-322) included several CVP-related sections.[2] These provisions directed pumping to "maximize" water supplies for the CVP (including pumping or "exports" to CVP water users south of the Sacramento and San Joaquin Rivers' confluence with the San Francisco Bay, known as the *Bay-Delta* or *Delta*) in accordance with applicable biological opinions (BiOps) for project operations.[3] They also allowed for increased pumping during certain storm events generating high flows, authorized actions to facilitate water transfers, and established a new standard for measuring the effects of water operations on species. In addition to operational provisions, the WIIN Act authorized funding for construction of new federal and nonfederal water storage projects. CVP projects are among the most likely recipients of this funding.

Due to increased precipitation and disagreements with the state, among other factors, the WIIN Act's CVP operational authorities did not yield significant new water exports south of the Delta in 2017 and 2018. However, the authorities may be more significant in years of limited precipitation and thus may yield increased supplies in the future. Although

---

[2] For more information, see CRS Report R44986, *Water Infrastructure Improvements for the Nation (WIIN) Act: Bureau of Reclamation and California Water Provisions*, by Charles V. Stern, Pervaze A. Sheikh, and Nicole T. Carter.

[3] The Endangered Species Act (ESA) requires that a federal agency proposing an action that may have an effect on a listed species consult with the U.S. Fish and Wildlife Service or the National Marine Fisheries Service (i.e., regulatory agencies). The action agency will commonly complete a biological assessment on potential effects to the fish or its habitat and submit it to the regulatory agency. The regulatory agency then renders a biological opinion, or BiOp, to the action agency making the proposal. The intent of a BiOp is to ensure that the proposed action will not reduce the likelihood of survival and recovery of an ESA-listed species. BiOps typically include conservation recommendations intended to further recovery of the ESA-listed species.

use of the new operational authorities was limited, Reclamation received funding for WIIN Act-authorized water storage project design and construction in FY2017-FY2019; a significant amount of this funding has gone to CVP-related projects.

Several state and federal proposals are also currently under consideration and have generated controversy for their potential to significantly alter CVP operations. In mid-2018, the State of California proposed revisions to its Bay-Delta Water Quality Control Plan. These changes would require that more flows from the San Joaquin and Sacramento Rivers reach the California Bay-Delta for water quality and fish and wildlife enhancement (and would thus further restrict water supplies for other users). At the same time, the Trump Administration is exploring options to increase CVP water supplies for users.

BACKGROUND

California's Central Valley encompasses almost 20,000 square miles in the center of the state (Figure 1). It is bound by the Cascade Range to the north, the Sierra Nevada to the east, the Tehachapi Mountains to the south, and the Coast Ranges and San Francisco Bay to the west. The northern third of the valley is drained by the Sacramento River, and the southern two-thirds of the valley are drained by the San Joaquin River. Historically, this area was home to significant fish and wildlife populations.

The CVP was originally conceived as a state project; the state studied the project as early as 1921, and the California state legislature formally authorized it for construction in 1933. After it became clear that the state was unable to finance the project, the federal government (through the U.S. Army Corps of Engineers, or USACE) assumed control of the CVP as a public works construction project authority provided under the Rivers and Harbors Act of 1935.[4] The Franklin D. Roosevelt Administration

---

[4] 49 Stat. 1028.

subsequently transferred the project to Reclamation.[5] Construction on the first unit of the CVP (Contra Costa Canal) began in October 1937, with water first delivered in 1940. Additional CVP units were completed and came online over time, and some USACE-constructed units have also been incorporated into the project.[6] The New Melones Unit was the last unit of the CVP to come online; it was completed in 1978 and began operations in 1979.

The CVP made significant changes to California's natural hydrology to develop water supplies for irrigated agriculture, municipalities, and hydropower, among other things. Most of the CVP's major units, however, predated major federal natural resources and environmental protection laws such as the Endangered Species Act (ESA; 87 Stat. 884. 16 U.S.C. §§1531-1544) and the National Environmental Policy Act (NEPA; 42 U.S.C. §§4321 et seq), among others. Thus, much of the current debate surrounding the project revolves around how to address the project's changes to California's hydrologic system that were not major considerations when it was constructed.

Today, CVP water serves a variety of different purposes for both human uses and fish and wildlife needs. The CVP provides a major source of support for California agriculture, which is first in the nation in terms of farm receipts.[7] CVP water supplies irrigate more than 3 million acres of land in central California and support 7 of California's top 10 agricultural counties. In addition, CVP M&I water provides supplies for approximately

---

[5] Transfer of the project to Reclamation was pursuant to a presidential directive in 1935 and subsequent congressional enactment of the Rivers and Harbors Act of 1937 (50 Stat. 844, 850).

[6] Although Reclamation constructed much of the Central Valley Project (CVP) and maintains control over its operations, the U.S. Army Corps of Engineers (USACE) has also been involved in the project over the course of its history. Some dams, such as Folsom Dam and New Melones Dam, initially were built by USACE but have been turned over to Reclamation for operations and maintenance and incorporated into the CVP. Additionally, USACE constructed and continues to operate several major dams in and around the Central Valley for flood control and other purposes, including Terminus Dam, Isabella Dam, Pine Flat Dam, and Success Dam in the San Joaquin Valley. Since USACE operates these dams for flood control, Reclamation administers contracts to use surplus water from these reservoirs for irrigation.

[7] U.S. Department of Agriculture, Economic Research Service, *Cash Receipts by State, Commodity Ranking and Share of U.S. Total, 2016,* at https://data.ers.usda.gov/reports.aspx?ID=17843#Pcb53fbff4c3c47c9b0afcc74d03a7403_3_17iT0R0x5.

2.5 million people per year. CVP operations are also critical for hydropower, recreation, and fish and wildlife protection. In addition to fisheries habitat, CVP flows support wetlands, which provide habitat for migrating birds.

Overview of the CVP and California Water Infrastructure

The CVP (Figure 1) is made up of 20 dams and reservoirs, 11 power plants, and 500 miles of canals, as well as numerous other conduits, tunnels, and storage and distribution facilities.[8] In an average year, it delivers approximately 5 million acre-feet (AF) of water to farms (including some of the nation's most valuable farmland); 600,000 AF to M&I users; 410,000 AF to wildlife refuges; and 800,000 AF for other fish and wildlife needs, among other purposes. A separate major project owned and operated by the State of California, the State Water Project (SWP), draws water from many of the same sources as the CVP and coordinates its operations with the CVP under several agreements. In contrast to the CVP, the SWP delivers about 70% of its water to urban users (including water for approximately 25 million users in the San Francisco Bay, Central Valley, and Southern California); the remaining 30% is used for irrigation.

At their confluence, the Sacramento and San Joaquin Rivers flow into the San Francisco Bay (the Bay-Delta, or Delta). Operation of the CVP and SWP occurs through the storage, pumping, and conveyance of significant volumes of water from both river basins (as well as trans-basin diversions from the Trinity River Basin in Northern California) for delivery to users. Federal and state pumping facilities in the Delta near Tracy, CA, export water from Northern California to Central and Southern California and are a hub for CVP operations and related debates. In the context of these controversies, *north of Delta* (NOD) and *south of Delta* (SOD) are important categorical distinctions for water users.

---

[8] Bureau of Reclamation, "About the Central Valley Project," at http://www.usbr.gov/mp/cvp/about-cvp.html.

CVP storage is spread throughout Northern and Central California. The largest CVP storage facility is Shasta Dam and Reservoir in Northern California (Figure 2), which has a capacity of 4.5 million AF. Other major storage facilities, from north to south, include Trinity Dam and Reservoir (2.4 million AF), Folsom Dam and Reservoir (977,000 AF), New Melones Dam and Reservoir (2.4 million AF), Friant Dam and Reservoir (520,000 AF), and San Luis Dam and Reservoir (1.8 million AF of storage, of which half is federal and half is nonfederal).

The CVP also includes numerous water conveyance facilities, the longest of which are the Delta-Mendota Canal (which runs for 117 miles from the federally operated Bill Jones pumping plant in the Bay-Delta to the San Joaquin River near Madera) and the Friant-Kern Canal (which runs 152 miles from Friant Dam to the Kern River near Bakersfield).

Non-CVP water storage and infrastructure is also spread throughout the Central Valley and in some cases is integrated with CVP operations. Major non-CVP storage infrastructure in the Central Valley includes multiple storage projects that are part of the SWP (the largest of which is Oroville Dam and Reservoir in Northern California), as well as private storage facilities (e.g., Don Pedro and Exchequer Dams and Reservoirs) and local government-owned dams and infrastructure (e.g., O'Shaughnessy Dam and Hetch-Hetchy Reservoir and Aqueduct, which are owned by the San Francisco Public Utilities Commission).

In addition to its importance for agricultural water supplies, California's Central Valley also provides valuable wetland habitat for migratory birds and other species. As such, it is home to multiple state, federal, and private wildlife refuges north and south of the Delta. Nineteen of these refuges (including 12 refuges within the National Wildlife Refuge system, 6 State Wildlife Areas/Units, and 1 privately managed complex) provide managed wetland habitat that receives water from the CVP and other sources. Five of these units are located in the Sacramento River Basin (i.e., North of the Delta), 12 are in the San Joaquin River Basin, and the remaining 2 are in the Tulare Lake Basin.[9]

---

[9] Tulare Lake, a freshwater dry lake in the San Joaquin River Valley, historically was one of the largest freshwater lakes west of the Great Lakes. Under most normal (nonflood) conditions,

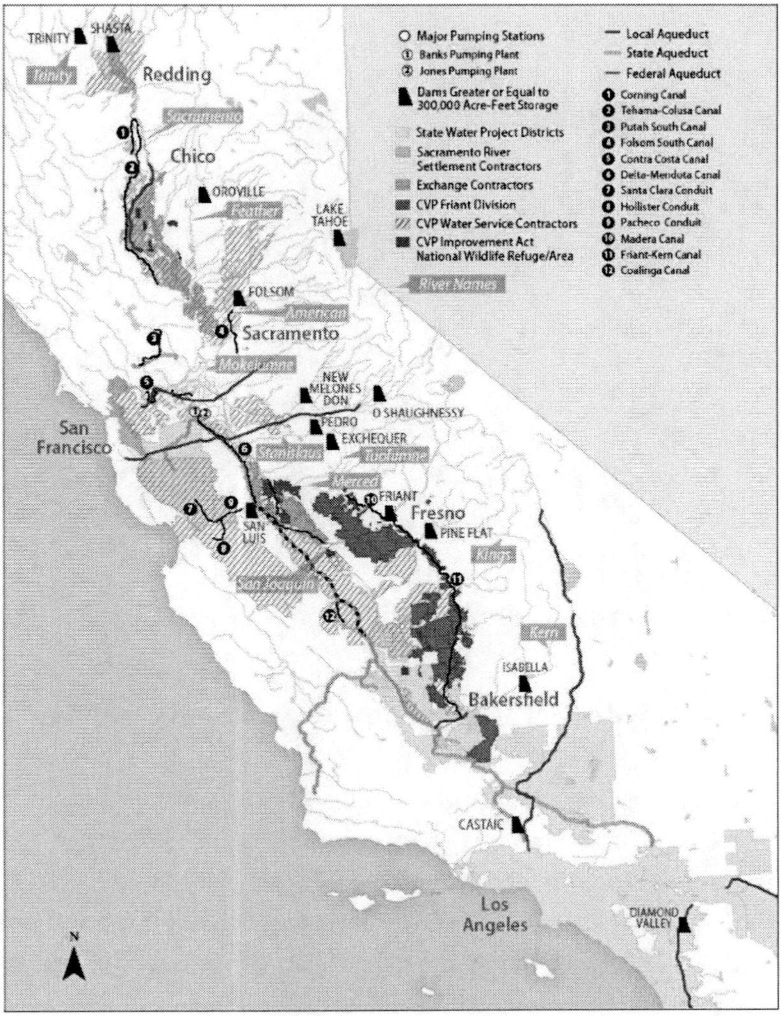

Source: Congressional Research Service (CRS).

Notes: Colored areas are based on water and irrigation district boundaries and do not correspond to the amount of water delivered from the Central Valley Project or the State Water Project. For example, some large areas have relatively small contracts for water compared with other, smaller areas.

Figure 1. Central Valley Project (CVP) and Related Facilities.

---

the lake was "terminal," meaning it had no outlet and did not drain downstream. Damming in the mid-20th century by the USACE of the Kaweah (Terminus Dam), Kern (Isabella Dam), Kings (Pine Flat Dam), and Tule Rivers (Success Dam), coupled with development of the basin for irrigated agriculture, dried up the lake bed under most conditions.

Source: Bureau of Reclamation.

Figure 2. Shasta Dam and Reservoir.

Central Valley Project Water Contractors and Allocations

In normal years, snowpack accounts for approximately 30% of California's water supplies and is an important factor in determining CVP and SWP allocations. Water from snowpack typically melts in the spring and early summer, and it is stored and made available to meet water needs throughout the state in the summer and fall. By late winter, the state's water supply outlook is typically sufficient for Reclamation to issue the amount of water it expects to deliver to its contractors.[10] At that time, Reclamation announces estimated deliveries for its 250 CVP water contractors in the upcoming water year.[11]

---

[10] A *water contractor*, as described in this chapter, has a contract for specified water deliveries from conveyance structures managed by the U.S. Bureau of Reclamation. Reclamation typically estimates these deliveries as a percentage of the total contract allocation to be made available for contractors within certain divisions, geographic areas, and/or contractor types (e.g., south-of-Delta agricultural contractors).

[11] A *water year* is a hydrologic unit for measuring a 12-month total for which precipitation totals are measured. In California, the water year typically is measured from October 1 of one year to September 30 of the following year.

Water Rights Contractors

California's system of state water rights has a profound effect on who gets how much water and when, particularly during times of drought or other restrictions on water supply. Because the waters of California are considered to be "the property of the people of the State," anyone wishing to use those waters must acquire a right to do so. California follows a dual system of water rights, recognizing both the riparian and prior appropriation doctrines. Under the riparian doctrine, a person who owns land that borders a watercourse has the right to make reasonable use of the water on that land (riparian rights). Riparian rights are reduced proportionally during times of shortage. Under the prior appropriation doctrine, a person who diverts water from a watercourse (regardless of his location relative thereto) and makes reasonable and beneficial use of the water acquires a right to that use of the water (appropriated rights). Appropriated rights are filled in order of seniority during times of shortage. Before exercising the right to use the water, appropriative users must obtain permission from the state through a permit system run by the State Water Resources Control Board (SWRCB).

Both the Central Valley Project (CVP) and the State Water Project (SWP) acquired rights for water use from the State of California, receiving several permits for water diversions at various points between 1927 and 1967. Since the Bureau of Reclamation found it necessary to take the water rights of other users to construct the CVP, it entered into settlement or exchange contracts with water users who had rights predating the CVP (and thus were senior users in time and right). Many of these special contracts were entered into in areas where water users were diverting water directly from the Sacramento and San Joaquin Rivers.

Sacramento River Settlement Contractors include the contractors (both individuals and districts) that diverted natural flows from the Sacramento River prior to the CVP's construction and executed a settlement agreement with Reclamation that provided for negotiated allocation of water rights. San Joaquin River Exchange Contractors are the irrigation districts that agreed to "exchange" exercising their water rights to divert water on the San Joaquin and Kings Rivers for guaranteed water deliveries from the CVP (typically in the form of deliveries from the Delta-Mendota Canal and waters north of the Delta). In contrast to water service contractors, water rights contractors receive 100% of their contracted amounts in most water-year types. During water shortages, their annual maximum entitlement may be reduced but not by more than 25%.

More than 9.5 million AF of water per year is *potentially* available from the CVP for delivery based on contracts between Reclamation and CVP contractors.[12] However, most CVP water contracts provide

[12] Water service contracts charge users a per-acre foot rate based on the amount of water delivered. In contrast, repayment contracts (the most common type of Reclamation contract

exceptions for Reclamation to reduce water deliveries due to hydrologic conditions and other conditions outside Reclamation's control.[13] As a result of these stipulations, Reclamation regularly makes cutbacks to actual CVP water deliveries to contractors due to drought and other factors.

Even under normal hydrological circumstances, the CVP often delivers much less than the maximum contracted amount of water; since the early 1980s, an average of about 7 million AF of water has been made available to CVP contractors annually (including 5 million AF to agricultural contractors). However, during drought years deliveries may be significantly less. In the extremely dry water years of 2012-2015, CVP annual deliveries averaged approximately 3.45 million AF.[14]

CVP contractors receive varying levels of priority for water deliveries based on their water rights and other related factors, and some of the largest and most prominent water contractors have a relatively low allocation priority. Major groups of CVP contractors include *water rights* contractors (i.e., senior water rights holders such as the Sacramento River Settlement and San Joaquin River Exchange Contractors, see box below), North and South of Delta water service contractors, and Central Valley refuge water contractors. The relative locations for these groups are shown in Figure 1.

The largest contract holders of CVP water by percentage of total contracted amounts are Sacramento River Settlement Contractors, located on the Sacramento River. The second-largest group are SOD water service contractors (including Westlands Water District, the CVP's largest contractor), located in the area south of the Delta. Other major contractors include San Joaquin River Exchange Contractors, located west of the San Joaquin River and Friant Division contractors, located on the east side of the San Joaquin Valley. Central Valley refuges and several smaller contractor groups (e.g., Eastside Contracts, In-Delta-Contra Costa

---

outside of the Central Valley Project [CVP]) charge users based on the amount of water storage allocated to a contractor, among other things.

[13] See U.S. Bureau of Reclamation, Mid-Pacific Region, *Final Form of Contract,4-19-2004*, Articles 3b, 11, 12a, and 12b, at http://www.usbr.gov/mp/cvpia/3404c/lt_contracts/index.html.

[14] CRS analysis of CVP contract water delivery information by the Bureau of Reclamation, October 3, 2018.

Contracts, and SOD Settlement Contracts) also factor into CVP water allocation discussions.[15] Figure 3 depicts an approximate division of *maximum* available CVP water deliveries pursuant to contracts with Reclamation. The largest contractor groups and their relative delivery priority are discussed in more detail in the Appendix to this chapter.

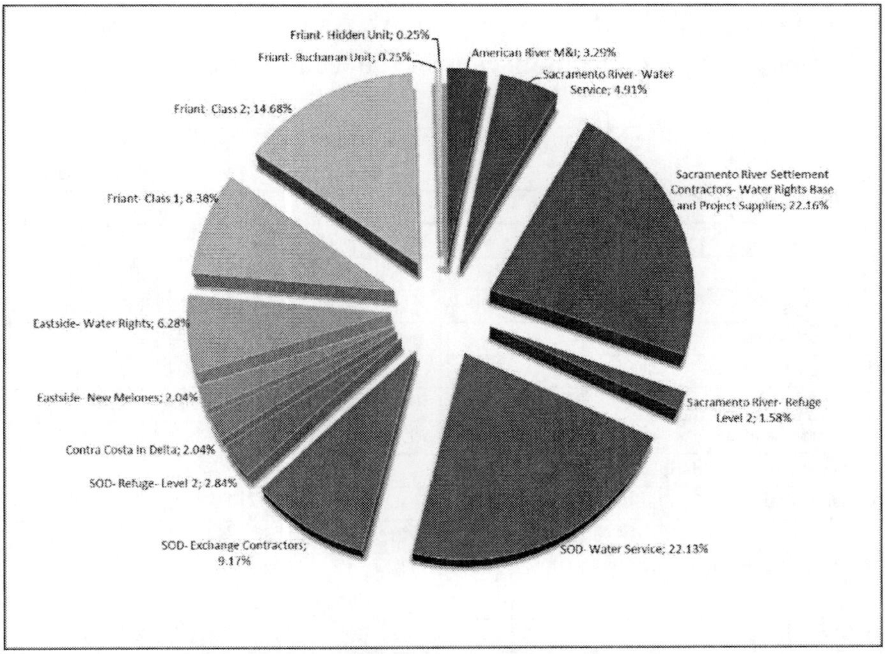

Source: CRS, using 2016 Bureau of Reclamation contractor data.

Notes: SOD = South-of-Delta; M&I = municipal and industrial water service contractors. Sacramento River Settlement Contractors includes both "base" water rights supplies (18.6%) and additional CVP "project" supplies (3.5%). For SOD Refuges, chart does not reflect "Level 4" supplies (for more information on Level 4 supplies, see below section, "Central Valley Wildlife Refuges").

Figure 3. Central Valley Project: Maximum Contract Amounts (relative share of total maximum contracted CVP supplies).

[15] Central Valley Project refuges are discussed more in the below section, "Central Valley Project Improvement Act."

CVP Allocations

Reclamation provided its allocations for the 2019 water year in a series of announcements in early 2019. As was the case in 2018, over the course of the spring Reclamation increased its allocations for some contractors from initially announced levels.

Table 1. CVP Water Allocations by Water Year, 2011-2019 (percentage of maximum contract allocation made available)

	2011	2012	2013	2014	2015	2016	2017	2018	2019 (est.)
North-of-Delta Users									
Agricultural	100%	100%	75%	0%	0%	100%	100%	100%	100%
M&I	100%	100%	100%	50%	25%	100%	100%	100%	100%
Settlement	100%	100%	100%	75%	75%	100%	100%	100%	100%
Contractors									
Refuges (Level 2)	100%	100%	100%	75%	75%	100%	100%	100%	100%
American River M&I	100%	100%	75%	50%	25%	100%	100%	100%	100%
In Delta- Contra	100%	100%	75%	50%	25%	100%	100%	100%	100%
Costa South-of-Delta Users									
Agricultural	80%	40%	20%	0%	0%	5%	100%	50%	75%
M&I	100%	75%	70%	50%	25%	55%	100%	70%	95%
Exchange Contractors	100%	100%	100%	65%	75%	100%	100%	100%	100%
Refuges (Level 2)	100%	100%	100%	65%	75%	100%	100%	100%	100%
Eastside Division	100%	100%	100%	55%	0%	0%	100%	100%	100%
Friant Class I	100%	50%	62%	0%	0%	65%	100%	88%	100%
Friant Class 2	20%	0%	0%	0%	0%	13%	100%	9%	10%a

Source: U.S. Bureau of Reclamation, CVP Historical Water Supply Allocations and 2019 Allocations, available at https://www.usbr.gov/mp/cvo/vungvari/water_allocations_historical.pdf.

Notes: CVP = Central Valley Project. M&I = municipal and industrial water service contractors. "Settlement" refers to contractors on the Sacramento River (north of Delta), and "Exchange" refers to contractors on the San Joaquin River (south of Delta) with special contracts and minimum delivery levels recognizing state water rights predating those acquired by the Bureau of Reclamation for construction and operation of the CVP. Contra Costa, Eastside Division, and Friant Class 1 and Class 2 represent individual or groups of water contractors.

[a] "Uncontrolled" Class 2 releases for Friant Contractors available through June 30, 2019.

Most CVP contractor groups were allocated 100% of their maximum contracted amounts in 2019. One major exception is SOD agricultural water service contractors, who were allocated 75% of their contracted supplies. Prior to receiving a full allocation in 2017, the last time these contractors received a 100% allocation was 2006. They have received their full contract allocations only four times since 1990.[16]

State Water Project Allocations

The other major water project serving California, the SWP, is operated by California's Department of Water Resources (DWR). The SWP primarily provides water to M&I users and some agricultural users, and it integrates its operations with the CVP. Similar to the CVP, the SWP has considerably more contracted supplies than it typically makes available in its deliveries. SWP contracted entitlements are 4.17 million AF, but average annual deliveries are typically considerably less than that amount.

SWP water deliveries were at their lowest point in 2014 and 2015, and they were significantly higher in the wet year of 2017. SWP water supply allocations for water years 2012-2019 are shown in Table 2.

Table 2. California State Water Project (SWP) Allocations by Water Year, 2012-2019 (percentage of maximum contract allocation)

	2012	2013	2014	2015	2016	2017	2018	2019 (est.)
State Water Project	65%	35%	5%	20%	60%	85%	35%	70%

Source: California Department of Water Resources, "Notices to State Water Project Contractors," https://water.ca.gov/Programs/State-Water-Project/Management/SWP-Water-Contractors.

[16] Full allocations were made in 1995, 1998, 2006, and 2017. Bureau of Reclamation, "Summary of Water Supply Allocations," at http://www.usbr.gov/mp/cvo/vungvari/water_allocations_historical.pdf.

Combined CVP/SWP Operations

The CVP and SWP are operated in conjunction under the 1986 Coordinated Operations Agreement (COA), which was executed pursuant to P.L. 99-546.[17] COA defines the rights and responsibilities of the CVP and SWP with respect to in-basin water needs and provides a mechanism to account for those rights and responsibilities. Despite several prior efforts to review and update the agreement to reflect major changes over time (e.g., water delivery reductions pursuant to the Central Valley Project Improvement Act, the Endangered Species Act requirements, and new Delta Water Quality Standards, among other things), the 1986 agreement remains in place.[18]

Combined CVP and SWP exports (i.e., water transferred from north to south of the Delta) is of interest to many observers because it reflects trends over time in the transfer of water from north to south (i.e., *exports*) by the two projects, in particular through pumping. Exports of the CVP and SWP, as well as total combined exports since 1978, have varied over time (Figure 4). Most recently, combined exports dropped significantly during the 2012-2016 drought but have rebounded since 2016. Prior to the drought, overall export levels had increased over time, having averaged more from 2001 to 2011 than over any previous 10-year period. The 6.42 million AF of combined exports in 2017 was the second most on record, behind 6.59 million AF in 2011.

Over time, CVP exports have decreased on average, whereas SWP exports have increased. Additionally, exports for agricultural purposes have declined as a subset of total exports, in part due to those exports being made available for other purposes (e.g., fish and wildlife).

---

[17] "Agreement Between the United States of America and the State of California for Coordinated Operation of the Central Valley Project and the State Water Project," No. 7-07-20-WO551. November 24, 1986.

[18] For more information on water delivery restrictions as they apply to the CVP, see below section, "Constraints on CVP Deliveries."

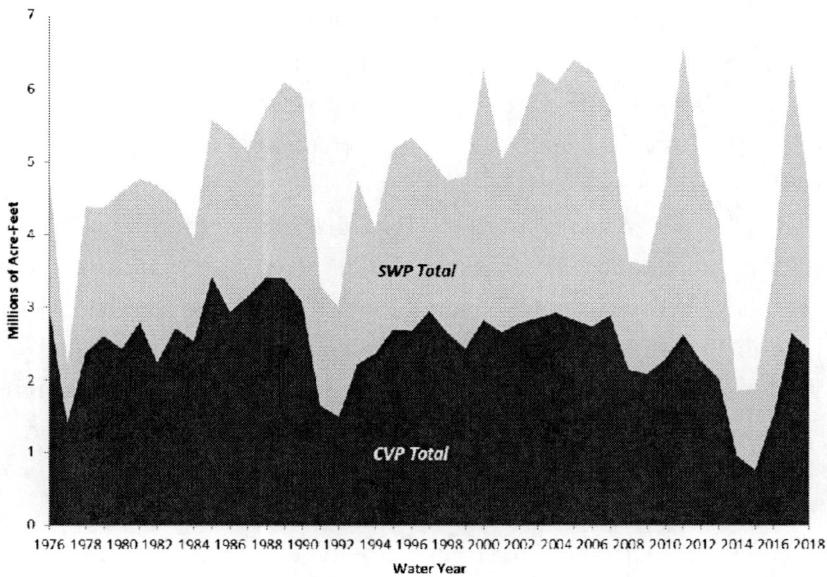

Source: CRS from data provided by the U.S. Dept. of the Interior, Bureau of Reclamation, email communication, June 19, 2019, *Total Annual Pumping at Banks, Jones, and Contra Costa Pumping Plants 1976-2018 (MAF).*

Figure 4. Central Valley Project (CVP) and State Water Project (SWP) Exports (exports in millions of acre-feet, 1978-2018).

Previously, some observers argued that CVP obligations under COA were no longer proportional to water supplies that the CVP receives from the Delta, thus the agreement should be renegotiated.[19] Dating to 2015, Reclamation and DWR conducted a mutual review of COA but were reportedly unable to agree on revisions. On August 17, 2018, Reclamation provided a Notice of Negotiations to DWR.[20] Following negotiations in the fall of 2018, Reclamation and DWR agreed to an addendum to COA in December 2018.[21] Whereas the original 1986 agreement included a fixed

[19] For example, see Joint Letter to the Bureau of Reclamation from Placer County Water Agency, City of Folsom, Tehama-Colusa Canal Authority et al, March 1, 2016, at http://www.ccwater.com/DocumentCenter/View/1854.

[20] Letter from David G. Murillo, Regional Directory, Bureau of Reclamation, to Karla Nemeth, Director, California Department of Water Resources, August 17, 2018.

[21] See Bureau of Reclamation and California Department of Water Resources, *Addendum to the Agreement Between the United States of America and the Department of Water Resources of the State of California for Coordinated Operation of the Central Valley Project and the State Water Project,* December 12, 2018.

ratio of 75% CVP/25% SWP for the sharing of regulatory requirements associated with storage withdrawals for Sacramento Valley in-basin uses (e.g., curtailments for water quality and species uses), the revised addendum adjusted the ratio of sharing percentages based on water year types (Table 3).

The 2018 addendum also adjusted the sharing of export capacity under constrained conditions. Whereas under the 1986 COA, export capacity was shared 50/50 between the CVP and the SWP, under the revised COA the split is to be 60% CVP/40% SWP during excess conditions, and 65% CVP/35% SWP during balanced conditions.[22] Finally, the state also agreed in the 2018 revisions to transport up to 195,000 AF of CVP water through the SWP's California Aqueduct, during certain conditions.

Table 3. COA Regulatory Requirements for CVP/SWP In-basin Storage Withdrawals (requirements pursuant to 1986 and 2018 agreements)

Water Year Type	1986 COA	COA with 2018 Addendum
All	75% CVP, 25% SWP	NA
Wet & Above Normal	NA	80% CVP, 20% SWP
Below Normal	NA	75% CVP, 25%SWP
Dry	NA	65% CVP, 35% SWP
Critically Dry	NA	60% CVP, 40% SWP

Source: Addendum to the Agreement between the United States of America and the Department of Water Resources of the State of California for Coordinated Operation of the Central Valley Project and the State Water Project, December 12, 2018.

CONSTRAINTS ON CVP DELIVERIES

Concerns over CVP water supply deliveries persist in part because even in years with high levels of precipitation and runoff, some contractors

[22] "Balanced" conditions refer to those conditions under which reservoir releases and unregulated flows in the Delta are equal to the water supply needed to meet Sacramento Valley-in-basin uses plus exports. Excess conditions are periods in which releases and unregulated flows exceed the aforementioned uses.

(in particular SOD water service contractors) have regularly received allocations of less than 100% of their contract supplies. Allocations for some users have declined over time; additional environmental requirements in recent decades have reduced water deliveries for human uses. Coupled with reduced water supplies available in drought years, some have increasingly focused on what can be done to increase water supplies for users. At the same time, others that depend on or advocate for the health of the San Francisco Bay and its tributaries, including fishing and environmental groups and water users throughout Northern California, have argued for maintaining or increasing existing environmental protections (the latter of which would likely further constrain CVP exports).

Hydrology and state water rights are the two primary drivers of CVP allocations. However, at least three other regulatory factors affect the timing and amount of water available for delivery to CVP contractors and are regularly the subject of controversy:

- State water quality requirements pursuant to state and the federal water quality laws (including the Clean Water Act [CWA, 33 U.S.C. §§1251-138]);
- Regulations and court orders pertaining to implementation of the federal Endangered Species Act (ESA, 87 Stat. 884. 16 U.S.C. §§1531-1544);[23] and
- Implementation of the Central Valley Project Improvement Act (CVPIA; P.L. 102-575).[24]

Each of these factors is discussed in more detail below.

---

[23] Requirements of the California Endangered Species Act (CESA) currently are being satisfied through implementation of the federal Endangered Species Act (ESA) due to a California state determination that project operations under the federal biological opinions are consistent with requirements under CESA. Presumably, if protections afforded to threatened and endangered species under the federal ESA were no longer in place, the State of California could invoke protections under CESA.

[24] P.L. 102-575, Title 34, 106 Stat. 4706.

Water Quality Requirements: Bay-Delta Water Quality Control Plan

California sets water quality standards and issues permits for the discharge of pollutants in compliance with the federal CWA, enacted in 1972.[25] Through the Porter-Cologne Act (a state law), California implements federal CWA requirements and authorizes the State Water Resources Control Board (State Water Board) to adopt water quality control plans, or basin plans.[26] The CVP and the SWP affect water quality in the Bay-Delta depending on how much freshwater the projects release into the area as "unimpaired flows" (thereby affecting area salinity levels).

The first Water Quality Control Plan for the Bay-Delta (Bay-Delta Plan) was issued by the State Water Board in 1978. Since then, there have been three substantive updates to the plan—in 1991, 1995, and 2006. The plans have generally required the SWP and CVP to meet certain water quality and flow objectives in the Delta to maintain desired salinity levels for in-Delta diversions (e.g., water quality levels for in-Delta water supplies) and fish and wildlife, among other things. These objectives often affect the amount and timing of water available to be pumped, or exported, from the Delta and thus at times result in reduced Delta exports to CVP and SWP water users south of the Delta.[27] The Bay-Delta Plan is currently implemented through the State Water Board's Decision 1641 (or D-1641), which was issued in 1999 and placed responsibility for plan

---

[25] The CWA requires the states to implement water quality standards that designate water uses to be protected and adopt water quality criteria that protect the designated uses. For application to California, see United States v. State Water Resources Control Board (Racanelli), 182 Cal. App. 3d 82, 109 (Cal. Ct. App. 1986).

[26] See Cal. Water Code §13160.

[27] Inability to reach agreement on water quality objectives through deliberation and litigation nearly shut down Delta pumping in the early 1990s and was a significant factor in the creation of the Bay-Delta Accord—a partnership between federal and state agencies with projects, responsibilities, and activities affecting the Delta. Habitat protection commitments in the accord were incorporated into the Bay-Delta Water Quality Control Plan, as were actions called for under the Vernalis Adaptive Management Program, and were included by the State Water Board in D-1641. (See U.S. Department of the Interior (DOI), Bureau of Reclamation, Mid-Pacific Region, *Long-Term Central Valley Project Operations Criteria and Plan*, Sacramento, CA, May 22, 2008, pp. 2-6).

implementation on the state's largest two water rights holders, Reclamation and the California DWR.[28]

Pumping restrictions to meet state-set water quality levels—particularly increases in salinity levels—can sometimes be significant. However, the relative magnitude of these effects varies depending on hydrology. For instance, Reclamation estimated that in 2014, water quality restrictions accounted for 176,300 AF of the reduction in pumping from the long-term average for CVP exports.[29] In 2016, Reclamation estimated that D-1641 requirements accounted for 114,500 AF in reductions from the long-term export average.

Bay-Delta Plan Update

In mid-2018, the State Water Board released the final draft of the update to the 2006 Bay Delta Plan (i.e., the Bay-Delta Plan Update) for the Lower San Joaquin River and Southern Delta. It also announced further progress on related efforts under the update for flow requirements on the Sacramento River and its tributaries.[30] The Bay-Delta Plan Update requires additional flows to the ocean (generally referred to in these documents as "unimpaired flows") from the San Joaquin River and its tributaries (i.e., the Stanislaus, Tuolumne, and Merced Rivers). Under the proposal, the unimpaired flow requirement for the San Joaquin River would be 40% (within a range of 30%-50%); average unimpaired flows currently range from 21% to 40%.[31] The state estimates that the updated version of the plan would reduce water available for human use from the San Joaquin River and its tributaries by between 7% and 23%, on average (depending

[28] California Environmental Protection Agency, State Water Resources Control Board, "Revised Water Right Decision 1641," March 15, 2000. https://www.waterboards.ca.gov/waterrights/board_decisions/adopted_orders/decisions/d1600_d1649/wrd1641_1999dec29.pdf.

[29] Personal communication with the Bureau of Reclamation, October 15, 2015.

[30] For more information, see the State Water Resources Control Board Bay Delta Plan update website at https://www.waterboards.ca.gov/waterrights/water_issues/programs/bay_delta/.

[31] California Water Boards, "State Water Board Seeks Public Comment on Final Draft Bay-Delta Plan Update for the Lower San Joaquin River and Southern Delta," July 6, 2018, at https://www.waterboards.ca.gov/waterrights/water_issues/programs/bay_delta/docs/Bay-Delta_Plan_Update_Press_Release.pdf.

on the water year type), but it could reduce these water supplies by as much as 38% during critically dry years.[32]

A more detailed plan for the Sacramento River and its tributaries is also expected in the future. A preliminary framework released by the state in July 2018 proposed a potential requirement of 55% unimpaired flows from the Sacramento River (within a range of 45% to 65%).[33] According to the State Water Board, if the plan updates for the San Joaquin and Sacramento Rivers are finalized and water users do not enter into voluntary agreements to implement them, the board could take actions to require their implementation, such as promulgation of regulations and conditioning of water rights.[34]

Reclamation and its contractors would likely play key roles in implementing any update to the Bay-Delta Plan, as they do in implementing the current plan under D-1641. Pursuant to Section 8 of the Reclamation Act of 1902,[35] Reclamation generally defers to state water law in carrying out its authorities, but the proposed Bay Delta Plan Update has generated controversy. In a July 2018 letter to the State Water Board, the Commissioner of Reclamation opposed the proposed standards for the San Joaquin River, arguing that meeting them would necessitate decreased water in storage at New Melones Reservoir of approximately 315,000 AF per year (a higher amount than estimated by the State Water Board). Reclamation argued that such a change would be contrary to the CVP prioritization scheme as established by Congress.[36]

---

[32] California Water Boards, "Summary of Proposed Amendments to the Bay-Delta Water Quality Control Plan," July 6, 2018, at https://www.waterboards.ca.gov/waterrights/water_issues/programs/bay_delta/docs/sed/lsjr_sdwq_summary_070618.pdf.

[33] California Water Boards, "July 2018 Framework for the Sacramento/Delta Update to the Bay-Delta Plan," July 6, 2018, at https://www.waterboards.ca.gov/waterrights/water_issues/programs/bay_delta/docs/sed/sac_delta_framework_070618%20.pdf. Hereinafter California Water Boards, "July 2018 Framework."

[34] California Water Boards, "July 2018 Framework."

[35] 43 U.S.C. §383.

[36] Letter from Brenda Burman, Commissioner, Bureau of Reclamation, DOI, to Felicia Marcus, Chair, State Water Resources Control Board, July 27, 2018. https://www.waterboards.ca.gov/waterrights/water_issues/programs/bay_delta/docs/comments_lsjr_finalsed/Brenda_Burman_BOR.pdf. Hereinafter Letter from Brenda Burman to Felicia Marcus.

On December 12, 2018, the State Water Board approved the Bay Delta Plan Update in Resolution 1018-0059.[37] According to the state, the plan establishes a "starting point" for increased river flows but also makes allowances for reduced river flows on tributaries where stakeholders have reached voluntary agreements to pursue both flow and "non-flow" measures.[38] The conditions in the Bay-Delta Plan Update would be implemented through water rights conditions imposed by the State Water Board; these conditions are to be implemented no later than 2022.

On March 28, 2019, the Department of Justice and DOI filed civil actions in federal and state court against the State Water Board for failing to comply with the California Environmental Quality Act.[39]

Endangered Species Act

Several species that have been listed under the federal ESA are affected by the operations of the CVP and the SWP.[40] One species, the Delta smelt, is a small pelagic fish that is susceptible to entrainment in CVP and SWP pumps in the Delta; it was listed as threatened under ESA in 1993. Surveys of Delta smelt in 2017 found two adult smelt, the lowest catch in the history of the survey.[41] These results were despite the

---

[37] California State Water Resources Control Board, Resolution No. 1018-0059, Adoption of Amendments to the Water Quality Control Plan for the San Francisco Bay/Sacramento-San Joaquin Delta Estuary and Final Substitute Environmental Document, December 12, 2018, https://www.waterboards.ca.gov/board_decisions/adopted_orders/resolutions/2018/rs2018_0059.pdf.

[38] California Water Boards, "State Water Board Adopts Bay-Delta Plan Update," press release, December 12, 2018, https://www.waterboards.ca.gov/press_room/press_releases/2018/pr121218_bay-delta_plan_update.pdf.

[39] Department of Justice, "United States Files Lawsuit Against California State Water Resources Control Board for Failure to Comply With California Environmental Quality Act," press release, March 28, 2019, https://www.justice.gov/opa/pr/united-states-files-lawsuit-against-california-state-water-resources-control-boardfailure.

[40] Act of December 28, 1973, P.L. 93-205; 87 Stat. 884, codified at 16 U.S.C. §§1531 et seq. This chapter assumes a basic knowledge of the act; an overview of the ESA and its major provisions may be found in CRS Report RL31654, *The Endangered Species Act: A Primer*, by Pervaze A. Sheikh.

[41] California Department of Fish and Wildlife, *Fall Midwater Trawl Monthly Abundance Index for Delta Smelt,* at http://www.dfg.ca.gov/delta/data/fmwt/indices.asp, accessed August 2, 2018.

relatively wet winter of 2017, which is a concern for many stakeholders because low population sizes of Delta smelt could result in greater restrictions on water flowing to users. It also raises larger concerns about the overall health and resilience of the Bay-Delta ecosystem. In addition to Delta smelt, multiple anadromous salmonid species are listed under ESA, including the endangered Sacramento River winter-run Chinook salmon, the threatened Central Valley spring-run Chinook salmon, the threatened Central Valley steelhead, threatened Southern Oregon/Northern California Coast coho salmon, and the threatened Central California Coast steelhead.

Federal agencies consult with the U.S. Fish and Wildlife Service (FWS) in DOI or the Department of Commerce's (DOC's) National Marine Fisheries Service (NMFS) to determine if a federal project or action might jeopardize the continued existence of a species listed under ESA or adversely modify its habitat. If an effect is possible, formal consultation is started and usually concludes with the appropriate service issuing a BiOp on the potential harm the project poses and, if necessary, issuing reasonable and prudent measures to reduce the harm.

FWS and NMFS each have issued federal BiOps on the coordinated operation of the CVP and the SWP. In addition, both agencies have undertaken formal consultation on proposed changes in the operations and have concluded that the changes, including increased pumping from the Delta, would jeopardize the continued existence of several species protected under ESA. To avoid such jeopardy, the FWS and NMFS BiOps have included Reasonable and Prudent Alternatives (RPAs) for project operations.

CVP and SWP BiOps have been challenged and revised over time. Until 2004, a 1993 winter-run Chinook salmon BiOp and a 1995 Delta smelt BiOp (as amended) governed Delta exports for federal ESA purposes. In 2004, a proposed change in coordinated operation of the SWP and CVP (including increased Delta exports), known as OCAP (Operations Criteria and Plan) resulted in the development of new BiOps. Environmental groups challenged the agencies' 2004 BiOps; this challenge resulted in the development of new BiOps by the FWS and NMFS in 2008 and 2009, respectively. These BiOps placed additional restrictions on the amount of water exported via SWP and CVP Delta pumps and other limitations on pumping and release of stored water.[42] The CVP and SWP are currently operated in accordance with these BiOps, both of which concluded that the coordinated long-term operation of the CVP and SWP, as proposed in Reclamation's 2008 Biological Assessment, was likely to jeopardize the continued existence of listed species and destroy or adversely modify designated critical habitat. Both BiOps included RPAs designed to allow the CVP and SWP to continue operating without causing jeopardy to listed species or destruction or adverse modification to designated critical habitat. Reclamation accepted and then began project operations consistent with the FWS and NMFS RPAs, which continue to govern operations.

The exact magnitude of reductions in pumping due to ESA restrictions compared to the aforementioned water quality restrictions has varied considerably over time. In absolute terms, ESA-driven reductions are typically greater in wet years than in dry years, but the proportion of ESA reductions relative to deliveries is not necessarily constant and depends on numerous factors. For instance, Reclamation estimated that ESA restrictions accounted for a reduction in deliveries of 62,000 AF from the long-term average for CVP deliveries in 2014 and 144,800 AF of CVP delivery reductions in 2015 (both years were extremely dry). In 2016, ESA reductions accounted for a much larger amount (528,000 AF) in a wet year, when more water is delivered.[43] Some scientists estimate that flows used to protect all species listed under ESA accounted for approximately 6.5% of the total Delta outflow from 2011 to 2016.[44]

During the 2012-2016 drought, implementation of the RPAs (which generally limit pumping under specific circumstances and call for water releases from key reservoirs to support listed species) was modified due to

---

[42] Among other things, the 2009 National Marine Fisheries Service biological opinion requires temperature considerations for the benefit of species in the Sacramento River and in the Bay-Delta. Operations of Shasta Dam and related facilities are thus affected by a separate plan, the Sacramento River Temperature Management Plan.

[43] Bureau of Reclamation, *Water Year 2016 CVIPA §3406(b)(2) Accounting*, at https://www.usbr.gov/mp/cvo/vungvari/FINAL_wy16_b2_800TAF_table_20170930.pdf.

[44] Peter B. Moyle, James A. Hobbs, and John R. Durand, "Delta Smelt and Water Politics in California," *Fisheries Magazine*, vol. 43, no. 1 (January 2018), pp. 42-60.

temporary urgency change orders (TUCs). These TUCs, issued by the State Water Resources Control Board in 2014 and again in 2015, were deemed consistent with the existing BiOps by NMFS and FWS. Such changes allowed more water to be pumped during certain periods based on real-time monitoring of species and water conditions. DWR estimates that approximately 400,000 AF of water was made available in 2014 for export due to these orders.[45]

In August 2016, Reclamation and DWR requested reinitiation of consultation on long-term, system-wide operations of the CVP and the SWP based on new information related to multiple years of drought, species decline, and related data.[46] In December 2017, the Trump Administration gave formal notice of its intent to prepare an environmental impact statement analyzing potential long-term modifications to the coordinated operations of the CVP and the SWP.[47] According to the notice, the actions under consideration will include those with the potential to "maximize" water and power supplies for users and that modify existing regulatory requirements, among other things.[48] The effort is widely viewed as an initial step toward potential long-term changes to CVP operations and existing BiOp requirements.

The Biological Assessment (BA) proposing changes for the operation of the CVP and SWP was sent to FWS and NMFS by Reclamation on

---

[45] California Environmental Protection Agency and State Water Resources Control Board, March 5, 2015 Order Modifying an Order That Approved in Part and Denied in Part a Petition for Temporary Urgency Changes to Permit Terms and Conditions Requiring Compliance with Delta Water Quality Objectives in Response to Drought Conditions, p. 4, at http://www.waterboards.ca.gov/waterrights/water_issues/programs/drought/docs/tucp/tucp_order030515.pdf.

[46] Letter from David Murillo, Regional Director, Bureau of Reclamation, and Mark W. Cowin, Director, Department of Water Resources, to Ren Lohoefener, Pacific Southwest Regional Director, August 2, 2016, at https://www.fws.gov/ sfbaydelta/documents/08-02-2016_BOR-DWR_Reinitiation_Letter.pdf.

[47] Bureau of Reclamation, "Notice of Intent to Prepare a Draft Environmental Impact Statement, Revisions to the Coordinated Long-Term Operation of the Central Valley Project and State Water Project, and Related Facilities," 82 *Federal Register* 61789-61791, December 29, 2017. Hereinafter Reclamation, "Intent to Prepare a Draft Environmental Impact Statement."

[48] Reclamation, "Intent to Prepare a Draft Environmental Impact Statement."

January 31, 2019.[49] The BA discusses the operational changes proposed by Reclamation and mitigation factors to address listed species. The changes reflect provisions in the WIIN Act and efforts to maximize water supplies for users. The BA also states that nonoperational activities will be implemented to augment and bolster listed fish populations. These activities include habitat restoration and introducing hatchery-bred Delta smelt. Operational changes include increasing flows to take into account additional water from winter storms and increasing base flows when storage levels are higher.

The Trump Administration has also indicated its intent to expedite other regulatory changes under ESA. On October 19, 2018, President Trump issued a memorandum that directed DOI and DOC to identify water infrastructure projects in California for which they have responsibilities under ESA. Per the memorandum, the agencies are to identify regulations and procedures that *burden* the projects and develop a plan to "suspend, revise, or rescind" those regulations.[50] The White House memorandum also directed that the aforementioned joint BiOps be completed by June 15, 2019.[51]

Central Valley Project Improvement Act

In an effort to mitigate many of the environmental effects of the CVP, Congress in 1992 passed the CVPIA as Title 34 of P.L. 102-575. The act made major changes to the management of the CVP. Among other things, it formally established fish and wildlife purposes as an official project purpose of the CVP and called for a number of actions to protect, restore, and enhance these resources. Overall, the CVPIA's provisions resulted in a combination of decreased water availability and increased costs for

---

[49] Bureau of Reclamation, *Updates to the Coordinated Long-Term Operation of the CVP and SWP and Related Facilities*, January 2019, https://www.usbr.gov/mp/bdo/lto.html.

[50] White House, "Presidential Memorandum on Promoting the Reliable Supply and Delivery of Water in the West," October 19, 2018. https://www.whitehouse.gov/presidential-actions/presidential-memorandum-promoting-reliablesupply-delivery-water-west/. Hereinafter, 2018 White House Memo on Western Water.

[51] 2018 White House Memo on Western Water.

agricultural and M&I contractors, along with new water and funding sources to restore fish and wildlife. Thus, the law remains a source of tension, and some would prefer to see it repealed in part or in full.

Some of the CVPIA's most prominent changes to the CVP included directives to

- double certain anadromous fish populations by 2002 (which did occur);[52]
- allocate 800,000 AF of "(b)(2)" CVP yield (600,000 AF in drought years) to fish and wildlife purposes;[53]
- provide water supplies (in the form of "Level 2" and "Level 4" supplies) for 19 designated Central Valley wildlife refuges;[54]
- establish a fund, the Central Valley Project Restoration Fund (CVPRF), to be financed by water and power users for habitat restoration and land and water acquisitions.

Pursuant to prior court rulings since enactment of the legislation, CVPIA (b)(2) allocations may be used to meet other state and federal requirements that reduce exports or require an increase from baseline reservoir releases. Thus, in a given year, the aforementioned export reductions due to state water quality and federal ESA restrictions are counted and reported on annually as (b)(2) water, and in some cases overlap with other stated purposes of CVPIA (e.g., anadromous fish restoration). The exact makeup of (b)(2) water in a given year typically varies. For example, in 2014 (a critically dry year), out of a total of 402,000 AF of (b)(2) water, 176,300 AF (44%) was attributed to export

---

[52] CVPIA's "fish doubling" goal was established on a baseline of average population levels during the period of 1967 1991.

[53] "*(b)(2) water*" references the provision in CVPIA that required these allocations.

[54] Authorized refuge water supply under CVPIA is divided into two categories: Level 2 and Level 4 supplies. Level 2 supplies (approximately 422,251 AF, except in critically dry years, when the allocation is reduced to 75%) are the historical average of water deliveries to the refuges prior to enactment of CVPIA. Reclamation is obligated to acquire and deliver this water under CVPIA, and costs are 100% reimbursable by CVP contractors through the Central Valley Project Restoration Fund. For more information, see **Appendix**.

reductions for Bay-Delta Plan water quality requirements.[55] Remaining (b)(2) water was comprised of a combination of reservoir releases classified as CVPIA anadromous fish restoration and NMFS BiOp compliance purposes (163,500 AF) and export reductions under the 2009 salmonid BiOp (62,200 AF).[56] In 2016 (a wet year), 793,000 AF of (b)(2) water included 528,000 AF (66%) of export pumping reductions under FWS and NMFS BiOps and 114,500 AF (14%) for Bay-Delta Plan requirements. The remaining water was accounted for as reservoir releases for the anadromous fish restoration programs, the NMFS BiOp, and the Bay-Delta Plan.[57]

ECOSYSTEM RESTORATION EFFORTS

Development of the CVP made significant changes to California's natural hydrology. In addition to the aforementioned CVPIA efforts to address some of these impacts, three ongoing, congressionally authorized restoration initiatives also factor into federal activities associated with the CVP:

- The Trinity River Restoration Program (TRRP), administered by Reclamation, attempts to mitigate impacts and restore fisheries impacted by construction of the Trinity River Division of the CVP.
- The San Joaquin River Restoration Program (SJRRP) is an ongoing effort to implement a congressionally enacted settlement to restore fisheries in the San Joaquin River.

---

55 Bureau of Reclamation, *Water Year 2014 CVPIA Section 3406(b)(2) Operations and Accounting*, January 28, 2015, at https://www.usbr.gov/mp/cvo/data/FINAL_WY14_b2_presentation.pdf.

56 Bureau of Reclamation, *Water Year 2014 CVPIA Section 3406(b)(2) Operations and Accounting*, January 28, 2015, at https://www.usbr.gov/mp/cvo/data/FINAL_WY14_b2_presentation.pdf.

57 Bureau of Reclamation, *Water Year 2016 CVIPA §3406(b)(2) Accounting*, at https://www.usbr.gov/mp/cvo/vungvari/FINAL_wy16_b2_800TAF_table_20170930.pdf.

- The California Bay-Delta Restoration Program aims to restore and protect areas within the Bay-Delta that are affected by the CVP and other activities.

In addition to their habitat restoration activities, both the TRRP and the SJRRP involve the maintenance of instream flow levels that use water that was at one time diverted for other uses. Each effort is discussed briefly below.

Trinity River Restoration Program

TRRP—administered by DOI—aims to mitigate impacts of the Trinity Division of the CVP and restore fisheries to their levels prior to the Bureau of Reclamation's construction of this division in 1955. The Trinity Division primarily consists of two dams (Trinity and Lewiston Dams), related power facilities, and a series of tunnels (including the 10.7-mile tunnel Clear Creek Tunnel) that divert water from the Trinity River Basin to the Sacramento River Basin and Whiskeytown Reservoir. Diversion of Trinity River water (which originally required that a minimum of 120,000 AF be reserved for Trinity River flows) resulted in the near drying of the Trinity River in some years, thereby damaging spawning habitat and severely depleting salmon stocks.

Efforts to mitigate the effects of the Trinity Division date back to the early 1980s, when DOI initiated efforts to study the issue and increase Trinity River flows for fisheries. Congress authorized legislation in 1984 (P.L. 98-541) and in 1992 (P.L. 102-575) providing for restoration activities and construction of a fish hatchery, and directed that 340,000 AF per year be reserved for Trinity River flows (a significant increase from the original amount). Congress also mandated completion of a flow evaluation study, which was formalized in a 2000 record of decision (ROD) that called for additional water for instream flows,[58] river channel restoration, and watershed rehabilitation.[59]

The 2000 ROD forms the basis for TRRP. The flow releases outlined in that document have in some years been supplemented to protect fish health in the river, and these increases have been controversial among some water users. From FY2013 to FY2018, TRRP was funded at approximately $12 million per year in discretionary appropriations from Reclamation's Fish and Wildlife Management and Development activity.

San Joaquin River Restoration Program

Historically, the San Joaquin River supported large Chinook salmon populations. After the Bureau of Reclamation completed Friant Dam on the San Joaquin River in the late 1940s, much of the river's water was diverted for agricultural uses and approximately 60 miles of the river became dry in most years. These conditions made it impossible to support Chinook salmon populations upstream of the Merced River confluence.

In 1988, a coalition of environmental, conservation, and fishing groups advocating for river restoration to support Chinook salmon recovery sued the Bureau of Reclamation. A U.S. District Court judge eventually ruled that operation of Friant Dam was violating state law because of its destruction of downstream fisheries.[60] Faced with mounting legal fees, considerable uncertainty, and the possibility of dramatic cuts to water diversions, the parties agreed to negotiate a settlement instead of proceeding to trial on a remedy regarding the court's ruling. This settlement was agreed to in 2006 and enacted by Congress in 2010 (Title X of P.L. 111-11).

---

[58] The additional flows outlined in the 2000 record of decision are based on water-year type and range from 369,000 AF in critically dry years to 815,000 AF in extremely wet years. A greater proportion of Trinity River water goes to the river in dry years, and a greater proportion of the water goes to CVP contractors in wet years.

[59] DOI, Record of Decision for Trinity River Mainstem Fishery Restoration Final Environmental Impact Statement/Environmental Impact Report, December 2000, at http://www.restoresjr.net/?wpfb_dl=2163.

[60] NRDC v. Patterson, 333 F. Supp. 2d 906, 925 (E.D. Cal. 2004).

The settlement agreement and its implementing legislation form the basis for the SJRRP, which requires new releases of CVP water from Friant Dam to restore fisheries (including salmon fisheries) in the San Joaquin River below Friant Dam (which forms Millerton Lake) to the confluence with the Merced River (i.e., 60 miles). The SJRRP also requires efforts to mitigate water supply delivery losses due to these releases, among other things. In combination with the new releases, the settlement's goals are to be achieved through a combination of channel and structural modifications along the San Joaquin River and the reintroduction of Chinook salmon (Figure 5). These activities are funded in part by federal discretionary appropriations and in part by repayment and surcharges paid by CVP Friant water users that are redirected toward the SJRRP in P.L. 111-11.

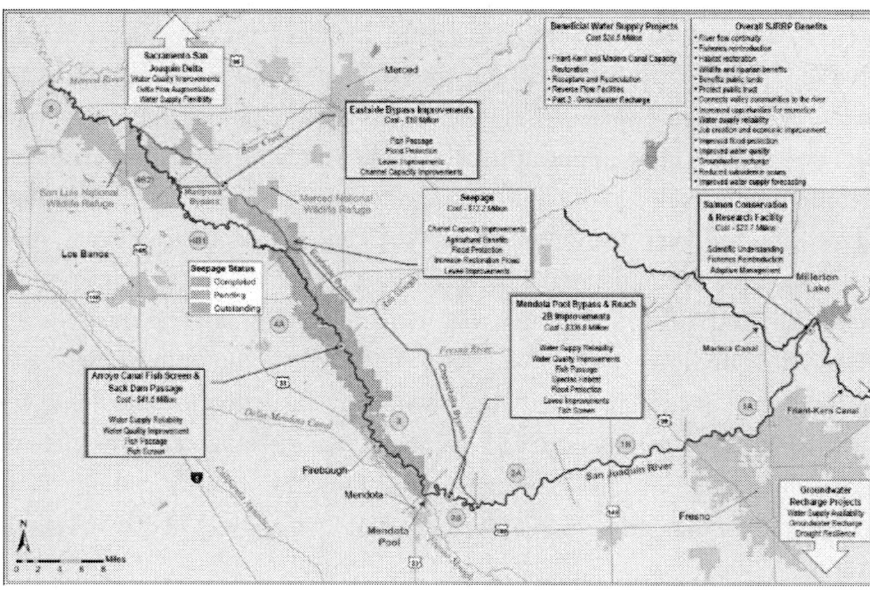

Source: Bureau of Reclamation, San Joaquin River Restoration Program, May 2018, at http://www.restoresjr.net/ ?wpfb_dl=2131.

Figure 5. San Joaquin River Restoration Program: Costs, Benefits, and Project Status (program details as of May 2018).

Because increased water flows for restoring fisheries (known as restoration flows) would reduce CVP diversions of water for off-stream purposes, such as irrigation, hydropower, and M&I uses, the settlement and its implementation have been controversial. The quantity of water used for restoration flows and the quantity by which water deliveries would be reduced are related, but the relationship is not necessarily one-for-one, due to flood flows in some years and other mitigating factors. Under the settlement agreement, no water would be released for restoration purposes in the driest of years; thus, the agreement would not reduce deliveries to Friant contractors in those years. Additionally, in some years, the restoration flows released in late winter and early spring may free up space for additional runoff storage in Millerton Lake, potentially minimizing reductions in deliveries later in the year—assuming Millerton Lake storage is replenished. Consequently, how deliveries to Friant water contractors may be reduced in any given year is likely to depend on many factors. Regardless of the specifics of how much water may be released for fisheries restoration vis-à-vis diverted for off-stream purposes, the SJRRP will impact existing surface and groundwater supplies in and around the Friant Division service area and affect local economies. SJRRP construction activities are in the early stages, but planning efforts have targeted a completion date of 2024 for the first stage of construction efforts.[61]

CALFED Bay-Delta Restoration Program

The Bay-Delta Restoration Program is a cooperative effort among the federal government, the State of California, local governments, and water users to proactively address the water management and aquatic ecosystem needs of California's Central Valley. The CALFED Bay-Delta Restoration Act (P.L. 108-361), enacted in 2004, provided new and expanded federal authorities for six agencies related to the 2000 ROD for the CALFED Bay-

---

[61] For more information, see San Joaquin River Restoration Program (SJRRP), *Funding Constrained Framework for Implementation*, May 2018.

Delta Program's Programmatic Environmental Impact Statement.[62] These authorities were extended through FY2019 under the WIIN Act. The interim action plan for CALFED has four objectives: a renewed federal-state partnership, smarter water supply and use, habitat restoration, and drought and floodplain management.[63]

From FY2013 to FY2018, Reclamation funded its Bay-Delta restoration activities at approximately $37 million per year; the majority of this funding has gone for projects to address the degraded Bay-Delta ecosystem and includes federal activities under California WaterFix (see below section, "California WaterFix").[64] Other agencies receiving funding to carry out authorities under CALFED include DOI's U.S. Fish and Wildlife Service and U.S. Geological Survey; the Department of Agriculture's Natural Resources Conservation Service; the Department of Defense's Army Corps of Engineers; the Department of Commerce's National Oceanic and Atmospheric Administration; and the Environmental Protection Agency. Similar to Reclamation, these agencies report on CALFED expenditures that involve a combination of activities under "base" authorities and new authorities that were provided under the CALFED authorizing legislation. The annual CALFED crosscut budget records the funding for CALFED across all federal agencies. The budget is generally included in the Administration's budget request and contains CALFED programs, their authority, and requested funding. For FY2019, the Administration requested $474 million for CALFED activities. This figure is an increase from the FY2018 enacted level of $415 million.

---

[62] CALFED Bay-Delta Program, *Programmatic Record of Decision*, August 28, 2000, at http://www.calwater.ca.gov/ content/Documents/ROD8-28-00.pdf. Other key documents available at http://www.calwater.ca.gov/calfed/library/Archive_ROD.html.

[63] *Interim Federal Action Plan for the California Bay-Delta*, December 22, 2009, at http://www.doi.gov/news/doinews/upload/CAWaterWorkPlan.pdf.

[64] In addition to funding under its CALFED authorities, Reclamation counts funding under its other CVP restoration authorities (e.g., CVPIA, SJRRP) as CALFED activities in its annual reporting.

NEW STORAGE AND CONVEYANCE

Reductions in available water deliveries due to hydrological and regulatory factors have caused some stakeholders, legislators, and state and federal government officials to look at other methods of augmenting water supplies. In particular, proposals to build new or augmented CVP and/or SWP water storage projects have been of interest to some policymakers. Additionally, the State of California is pursuing a major water conveyance project, the California WaterFix, with a nexus to CVP operations.

New and Augmented Water Storage Projects

The aforementioned CALFED legislation (P.L. 108-361) also authorized the study of several new or augmented CVP storage projects throughout the Central Valley that have been ongoing for a number of years. These studies include Shasta Lake Water Resources Investigation, North of the Delta Offstream Storage Investigation (also known as Sites Reservoir), In-Delta Storage, Los Vaqueros Reservoir Expansion, and Upper San Joaquin River/Temperance Flat Storage Investigation (Figure 6). Although the recommendations of these studies would normally be subject to congressional approval, Section 4007 of the WIIN Act authorized $335 million in Reclamation financial support for new or expanded federal and nonfederal water storage projects and provided that these projects could be deemed authorized, subject to a finding by the Administration that individual projects met certain criteria.[65]

In 2018 reporting to Congress, Reclamation recommended an initial list of seven projects that it concluded met the WIIN Act criteria. The projects were allocated $33.3 million in FY2017 funding that was previously appropriated for WIIN Act Section 4007 projects. Congress approved the funding allocations for these projects in enacted

[65] For more information, see CRS In Focus IF10626, *Reclamation Water Storage Projects: Section 4007 of the Water Infrastructure Improvements for the Nation Act*, by Charles V. Stern.

appropriations for FY2018 (P.L. 115-141). Four of the projects receiving FY2017 funds ($28.05 million) were CALFED studies that would address water availability in the CVP:[66]

- Shasta Dam and Reservoir Enlargement Project ($20 million for design and preconstruction);
- North-of-Delta Off-Stream Storage Investigation/Sites Reservoir Storage Project ($4.35 million for feasibility study);
- Upper San Joaquin River Basin Storage Investigation ($1.5 million for feasibility study); and
- Friant-Kern Canal Subsidence Challenges Project ($2.2 million for feasibility study).

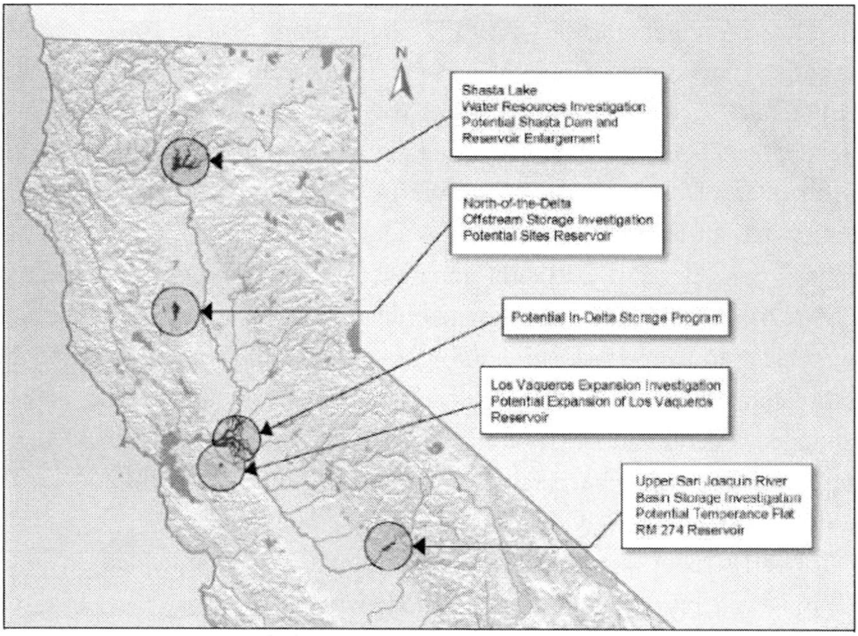

Source: California Department of Water Resources, A Resource Management Strategy of the California Water Plan, July 29, 2016.

Figure 6. CALFED Surface Water Storage Studies.

---

[66] The study of several projects (including the Shasta, Sites, and Upper San Joaquin investigations) was originally authorized under P.L. 108-361.

The enacted FY2018 Energy and Water appropriations bill further stipulated that $134 million of the amount set aside for additional water conservation and delivery projects be provided for Section 4007 WIIN Act storage projects (i.e., similar direction as FY2017). The enacted FY2019 bill set aside another $134 million for these purposes.[67] Future reporting and appropriations legislation is expected to propose allocation of this and any other applicable funding. Congress also may consider additional directives for these and other efforts to address water supplies in the CVP, including approval of physical construction for one or more of these projects.

Funding by the State of California may also influence the viability and timing of construction for some of the proposed projects. For example, in June 2018, the state announced significant bond funding for Sites Reservoir ($1.008 billion), as well as other projects.[68]

California WaterFix

In addition to water storage, some have advocated for a more flexible water conveyance system for CVP and SWP water. An alternative was the California WaterFix, a project initiated by the State of California in 2015 to address some of the water conveyance and ecosystem issues in the Bay-Delta. The objective of this project was to divert water from the Sacramento River, north of the Bay-Delta, into twin tunnels running south along the eastern portion of the Bay-Delta and emptying into existing pumps that feed water into the CVP and SWP. In the spring of 2019, Governor Newsom of California canceled the plans for this project and introduced an alternative plan for conveying water through the Delta.

DWR is creating plans to construct a single tunnel to convey water from the Sacramento River to the existing pumps in the Bay-Delta. DWR's

---

[67] See CRS In Focus IF10841, *Bureau of Reclamation: FY2019 Appropriations*, by Charles V. Stern.

[68] California Water Commission, "Commission Approves Investing $2.7 Billion in Eight Water Storage Projects," July 24, 2018, at https://cwc.ca.gov/-/media/CWC-Website/Files/Documents/Press/MCEDPressRelease_072418.pdf.

stated reasons for supporting this approach are to protect water supplies from sea-level rise, saltwater intrusion, and earthquakes.[69] The new plan is expected to take a "portfolio" approach that focuses on a number of interrelated efforts to make water supplies climate resilient. This approach includes actions such as strengthening levees, protecting Delta water quality, and recharging groundwater, according to DWR.[70] This project will require a new environmental review process for federal and state permits. It is being led by the Delta Conveyance Design and Construction Authority, a joint powers authority created by public water agencies to oversee the design and construction of the new conveyance system.[71] DWR is expected to oversee the planning effort. The cost of the project is anticipated to be largely paid by public water agencies. The federal government's role in this project beyond evaluating permit applications and maintaining related CVP operations has not been defined.

CONGRESSIONAL INTEREST

Congress plays a role in CVP water management and has previously attempted to make available additional water supplies in the region by facilitating efforts such as water banking, water transfers, and construction of new and augmented storage. In 2016, Congress enacted provisions aiming to benefit the CVP and the SWP, including major operational changes in the WIIN Act and additional appropriations for western drought response and new water storage that have benefited (or are expected to benefit) the CVP. Congress also continues to consider legislation that would further alter CVP operational authorities and responsibilities related

---

[69] California Department of Water Resources, "State Withdraws WaterFix Approvals, Initiates Planning and Permitting for a Smaller Single Tunnel," press release, May 2, 2019, https://water.ca.gov/News/News-Releases/2019/May/StateWithdraws-WaterFix-Approvals. Hereinafter "DWR May 2019 Press Release."

[70] DWR May 2019 Press Release.

[71] California Department of Water Resources, *Modernizing Delta Conveyance Infrastructure Q&A*, California Department of Water Resources, https://water.ca.gov/-/media/DWR-Website/Web-Pages/Programs/Delta-Conveyance/Delta-Conveyance-QA.pdf?la=en&hash=373E0DBCD7AD988C9987A3197304E55D9115F798.

to individual units of the project. The below section discusses some of the main issues related to the CVP that may receive attention by Congress.

CVP Operational Authorities under the WIIN Act[72]

Title II, Subtitle J of the WIIN Act (enacted in December 2016) included multiple provisions related to the Bureau of Reclamation's operations of the CVP. Most of the WIIN Act's operational provisions are set to expire in 2021 (five years after the bill's enactment). In addition to overseeing the implementation of these operational provisions, Congress may also consider their amendment, extension, or repeal.

The WIIN Act directed Reclamation to "maximize" CVP pumping (in accordance with applicable BiOps), allowed for increased pumping during certain temporary storm events, and authorized expedited reviews of water transfers, among other things.[73] The WIIN Act also established a new standard for measuring the effects of water operations on species listed as endangered or threatened under the ESA, allowing most of the bill's actions to go forward unless they are determined to cause additional adverse effects on listed species beyond the range of the effects anticipated to occur for the duration of the species BiOp.

Although the WIIN Act included some provisions from legislation that had been proposed dating back to the 112th Congress, many of the controversial provisions from prior bills were not included in the act. Supporters of WIIN Act operational changes contended that these changes had the potential to make additional water available to users facing curtailed deliveries, while also improving the flexibility and responsiveness of the management and operations of the CVP and SWP.[74]

---

[72] For more information on these provisions, see CRS Report R44986, *Water Infrastructure Improvements for the Nation (WIIN) Act: Bureau of Reclamation and California Water Provisions*, by Charles V. Stern, Pervaze A. Sheikh, and Nicole T. Carter.

[73] Pursuant to Section 4013 of the WIIN Act, most of the CVP operational provisions expire in December 2021.

[74] U.S. Congress, Senate Committee on Energy and Natural Resources, *Statement of Dan Keppen, Executive Director, Family Farm Alliance,* 114th Cong., 1st sess., October 8, 2015, p. 4, at http://www.energy.senate.gov/public/index.cfm/hearings-and-business-meetings?

Opponents worried that the changes may have detrimental effects on species' survival in both the short and long terms and may limit agency efforts to manage water supplies for the benefit of species.[75] Some of the notable CVP operational provisions in the WIIN Act aimed to provide the Administration with authority to make available more water supplies during periods in which pumping otherwise would have been limited.

According to Reclamation, some changes authorized under the WIIN Act were implemented during the winter of 2017-2018. In particular, communication and transparency were reportedly increased for some operational decisions, allowing for reduced or rescheduled pumping restrictions. Additionally, as of spring 2018, WIIN Act allowances relaxed restrictions on inflow-to-export ratios related to the voluntary sale, transfer, or exchange of water that were used to affect a transfer resulting in additional exports of 50,000-60,000 AF.[76]

Reclamation has noted that hydrology has affected its ability to implement some of the act's provisions. Many of the WIIN Act changes have the potential to make their greatest impact during drought years. At the same time, some federal operational changes pursuant to the WIIN Act were reportedly proposed but were deemed incompatible with state requirements.[77] Despite these limitations, WIIN Act authorities are likely to continue as a topic of congressional interest.

Other Proposed Changes to CVP Operations

Previous Congresses have considered legislation that proposed additional changes to CVP operations.[78] For instance, in the 115th

---

Id=65220e15-0479-492e-8423-ca1a381c1078&Statement_id=378db42f-6b60-44a7-a16c-3d2b7d712984.

[75] U.S. Congress, Senate Committee on Energy and Natural Resources, *Statement of Michael L. Connor, Deputy Secretary, U.S. Department of the Interior*, 114th Cong., 1st sess., October 8, 2015, p. 1, at http://www.energy.senate.gov/public/index.cfm/files/serve?File_id=fb299e7d-7de8-41c8-b8a2-365d544c8911.

[76] This provision of the WIIN Act generally lessened existing restrictions on the amount of water that could be exported for water transfers. Personal communication with the Bureau of Reclamation, April 18, 2017.

[77] Personal communication with the Bureau of Reclamation, April 18, 2017.

[78] For more information, see CRS Report R44889, *H.R. 23, the Gaining Responsibility on Water Act of 2017 (GROW Act)*, by Charles V. Stern, Pervaze A. Sheikh, and Marnie Kremer.

Congress, H.R. 23, the Gaining Responsibility on Water Act (GROW Act), incorporated a number of provisions that were included in previous California drought legislation in the 112th, 113th, and 114th Congresses but were not enacted in the WIIN Act. Generally speaking, the GROW Act included provisions that would have loosened some environmental protections and restrictions that are imposed under the CVPIA, ESA, CWA, and SJRRP, and had the potential to increase exports under some scenarios. This legislation was not enacted.

In addition to legislation proposing operational changes, the Administration has indicated its intent to propose administrative changes to CVP operations, including through reinitiation of consultation on long-term, system-wide operations of the CVP and SWP (see earlier section, "Endangered Species Act"). A 2018 White House memorandum directed DOC and DOI to finalize their new BiOps for the coordinated operation of the CVP and SWP by June 15, 2019, and to "suspend, revise, or rescind" regulations that unduly burden the project.[79] It is unclear how the latter process might unfold or what particular regulations will be addressed.

New Water Storage Projects

As previously noted, Reclamation and the State of California have funded the study of new water storage projects in recent years, and future appropriations legislation and reporting may provide additional direction for these and other efforts to develop new water supplies for the CVP. As such, Congress may consider oversight, authorization, and/or funding for these projects. Some projects, such as the Shasta Dam and Reservoir Enlargement Project, have the potential to augment CVP water supplies but have also generated controversy for their potential to conflict with the intent of certain state laws.[80] Although Reclamation has indicated its

---

[79] 2018 White House Memo on Western Water.

[80] In particular, Section 5093.542 of the California State Public Resources Code prevents participation (other than technical or economic feasibility studies of the Shasta dam raise project) by state departments or agencies in facilities that would have an adverse effect on

interest in pursuing the Shasta Dam raise project, the state has opposed the project under Governor Jerry Brown's Administration, and it is unclear how such a project might proceed absent state regulatory approvals and financial support. As previously noted, in early 2018, Reclamation proposed and Congress agreed to $20 million in design and preconstruction funding for the project.[81] An additional $75 million was recommended by the Trump Administration in February 2019.

In addition to the Shasta Dam and Reservoir Enlargement Project, Congress approved Reclamation-recommended study funding for Sites Reservoir/North of Delta Offstream Storage (NODOS), Upper San Joaquin River Basin Storage Investigation, and the Friant-Kern Canal Subsidence Challenges Project. Overall, from FY2017 to FY2019 Congress provided Reclamation with $335 million for new water storage projects authorized under Section 4007 of the WIIN Act. A significant share of this total is expected to be used on CVP and related water storage projects in California. Once the appropriations ceiling for these projects has been reached, funding for storage projects under Section 4007 would need to be extended by Congress before projects could proceed further.

Legislation in the 116th Congress has proposed to expedite certain water storage studies in the Central Valley, and could also provide mandatory funding for their eventual construction. For instance, Section 5 of H.R. 2473 would direct the Secretary to complete, as soon as practicable, the ongoing feasibility studies associated with Sites Reservoir, Del Puerto Canyon Reservoir, Los Vaqueros Reservoir, and San Luis Reservoir. Section 2 of the same legislation would authorize $100 million per year for fiscal years 2030 to 2060, without further appropriation (i.e., mandatory funding) for new Reclamation surface or groundwater storage projects.

---

the free-flowing condition of the McCloud River. In previous documents, Reclamation has indicated that this requirement could limit some state agency participation in the project.

[81] This funding was provided from a pool of funds appropriated for FY2017 that was designated for water storage projects authorized under Section 4007 of the WIIN Act. Enacted appropriations in FY2018 and FY2019 have included similar funding amounts. For more information, see CRS In Focus IF10692, *Bureau of Reclamation: FY2018 Appropriations*, by Charles V. Stern.

Conclusion

The CVP is one of the largest and most complex water storage and conveyance projects in the world. Congress has regularly expressed interest in CVP operations and allocations, in particular pumping in the Bay-Delta. In addition to ongoing oversight of project operations and previously enacted authorities, a number of developing issues and proposals related to the CVP have been of interest to congressional decisionmakers. These include study and approval of new water storage and conveyance projects, updates to the state's Bay-Delta Water Quality Plan, and a multipronged effort by the Trump Administration to make available more water for CVP water contractors, in particular those south of the Delta. Future drought or other stressors on California water supplies are likely to further magnify these issues.

Appendix. CVP Water Contractors

The below sections provide a brief discussion some of the major contractor groups and individual contractors served by the CVP.

Sacramento River Settlement Contractors and San Joaquin River Exchange Contractors (Water Rights Contractors)

CVP water is generally made available for delivery first to those contractors north and south of the Delta with water rights that predate construction of the CVP: the Sacramento River Settlement Contractors and the San Joaquin River Exchange Contractors. (These contractors are sometimes referred to collectively as water rights contractors.) Water rights contractors typically receive 100% of their contracted amounts in most water year types. During water shortages, their annual maximum entitlement may be reduced, but not by more than 25%.

Sacramento River Settlement Contractors include the 145 contractors (both individuals and districts) that diverted natural flows from the Sacramento River prior to the CVP's construction and executed a settlement agreement with Reclamation that provided for negotiated allocation of water rights. Reclamation entered into this agreement in exchange for these contractors withdrawing their protests related to Reclamation's application for water rights for the CVP.

The San Joaquin River Exchange Contractors are four irrigation districts that agreed to "exchange" exercising their water rights to divert water on the San Joaquin and Kings Rivers for guaranteed water deliveries from the CVP (typically in the form of deliveries from the Delta-Mendota Canal and waters north of the Delta). During all years except for when critical conditions are declared, Reclamation is responsible for delivering 840,000 AF of "substitute" water to these users (i.e., water from north of the Delta as a substitute for San Joaquin River water). In the event that Reclamation is unable to make its contracted deliveries, these Exchange Contractors have the right to divert water directly from the San Joaquin River, which may reduce water available for other San Joaquin River water service contactors.

Friant Division Contractors

CVP's Friant Division contractors receive water stored behind Friant Dam (completed in 1944) in Millerton Lake. This water is delivered through the Friant-Kern and Madera Canals. The 32 Friant Division contractors, who irrigate roughly 1 million acres on the San Joaquin River, are contracted to receive two "classes" of water: Class 1 water is the first 800,000 AF available for delivery;[82] Class 2 water is the next 1.4 million AF available for delivery. Some districts receive water from both classes. Generally, Class 2 waters are released as "uncontrolled flows" (i.e., for

[82] This water typically is provided for municipal and industrial use or for districts without access to groundwater.

flood control concerns), and may not necessarily be scheduled at a contractor's convenience.

Deliveries to the Friant Division are affected by a 2009 congressionally enacted settlement stemming from Friant Dam's effects on the San Joaquin River.[83] The settlement requires reductions in deliveries to Friant users for protection of fish and wildlife purposes. In some years, some of these "restorations flows" have been made available to contractors for delivery as Class 2 water.

Unlike most other CVP contractors, Friant Division contractors have converted their water service contracts to repayment contracts and have repaid their capital obligation to the federal government for the development of their facilities. In years in which Reclamation is unable to make contracted deliveries to Exchange Contractors, these contractors can make a "call" on water in the San Joaquin River, thereby requiring releases from Friant Dam that otherwise would go to Friant contractors.

South-of-Delta (SOD) Water Service Contractors: Westlands Water District

As shown in Figure 3, SOD water service contractors account for a large amount (2.09 million AF, or 22.1%) of the CVP's contracted water. The largest of these contractors is Westlands Water District, which consists of 700 farms covering more than 600,000 acres in Fresno and Kings Counties. In geographic terms, Westlands is the largest agricultural water district in the United States; its lands are valuable and productive, producing more than $1 billion of food and fiber annually.[84] Westlands' maximum contracted CVP water is in excess of 1.2 million AF, an amount that makes up more than half of the total amount of SOD CVP water

---

[83] When constructed, Friant Dam impounded the entire flow of the San Joaquin River, except for releases to manage flooding and provide water for some riparian water rights holders immediately below the dam. For more information, see below section, "San Joaquin River Restoration Program."

[84] Westlands Water District, "Who We Are," at https://wwd.ca.gov/wp-content/uploads/2016/08/wwd-who-we-are.pdf.

service contracts and significantly exceeds any other individual CVP contactor.[85] However, due to a number of factors, Westlands often receives considerably less water on average than it did historically.

Westlands has been prominently involved in a number of policy debates, including proposals to alter environmental requirements to increase pumping south of the Delta. Westlands is also involved in a major proposed settlement with Reclamation, the San Luis Drainage Settlement. The settlement would, among other things, forgive Westlands' share of federal CVP repayment responsibilities in exchange for relieving the federal government of its responsibility to construct drainage facilities to deal with toxic runoff associated with naturally occurring metals in area soils.

Central Valley Wildlife Refuges

The 20,000 square mile California Central Valley provides valuable wetland habitat for migratory birds and other species. As such, it is the home to multiple state and federally-designated wildlife refuges north and south of the Delta. These refuges provide managed wetland habitat that receives water from the CVP and other sources.

The Central Valley Project Improvement Act (CVPIA; P.L. 102-575),[86] enacted in 1992, sought to improve conditions for fish and wildlife in these areas by providing them coequal priority with other project purposes. CVPIA also authorized a Refuge Water Supply Program to acquire approximately 555,000 AF annually in water supplies for 19 Central Valley refuges administered by three managing agencies: California Department of Fish and Wildlife, U.S. Fish and Wildlife Service, and Grassland Water District (a private landowner). Pursuant to CVPIA, Reclamation entered into long-term water supply contracts with the managing agencies to provide these supplies.

---

[85] CRS analysis of Bureau of Reclamation, "Central Valley Project Water Contractors," March 30, 2016, at https://www.usbr.gov/mp/cvp-water/docs/latest-water-contractors.pdf.

[86] P.L. 102-575, Title 34, 106 Stat. 4706.

Authorized refuge water supply under CVPIA is divided into two categories: Level 2 and Level 4 supplies. Level 2 supplies (approximately 422,251 AF, except in critically dry years, when the allocation is reduced to 75%) are the historical average of water deliveries to the refuges prior to enactment of CVPIA.[87] Reclamation is obligated to acquire and deliver this water under CVPIA, and costs are 100% reimbursable by CVP contractors through a fund established by the act, the Central Valley Project Restoration Fund (CVPRF; see previous section, "Central Valley Project Improvement Act"). Level 4 supplies (approximately 133,264 AF) are the additional increment of water beyond Level 2 supplies for optimal wetland habitat development. This water must be acquired by Reclamation through voluntary measures and is funded as a 75% federal cost (through the CVPRF) and 25% state cost.

In most cases, the Level 2 requirement is met; however, Level 4 supplies have not always been provided in full for a number of reasons, including a dearth of supplies due to costs in excess of available CVPRF funding and a lack of willing sellers. In recent years, costs for the Refuge Water Supply Program (i.e., the costs for both Level 2 and Level 4 water) have ranged from $11 million to $20 million.

---

[87] Although this represents the historical average for deliveries, prior to the Central Valley Project Improvement Act (CVPIA; P.L. 102-575), refuges only had a legal entitlement to 121,700 acre-feet (AF).

In: Key Congressional Reports for June 2019 ISBN: 978-1-53616-662-0
Editor: Stephanie A. Betts

Chapter 4

BIOMASS: COMPARISON OF DEFINITIONS IN LEGISLATION*

Kelsi Bracmort

SUMMARY

The use of biomass as an energy feedstock has regularly been presented as a potentially viable alternative to address U.S. energy security concerns, foreign oil dependence, and rural economic development, and as a tool to possibly help improve the environment (e.g., through greenhouse gas emission reduction). Biomass (organic matter that can be converted into energy) includes food crops, crops grown specifically to produce energy (e.g., switchgrass or prairie perennials), crop residues, wood waste and byproducts, and animal manure. Biomass may be used to produce heat, electricity, or liquid transportation fuel. Efforts to promote the use of biomass for liquid transportation fuels have focused on corn primarily. Efforts to promote the use of biomass for power generation have focused on wood, wood residues, and milling waste primarily. Comparatively less emphasis has been placed on the use of other biomass feedstocks—noncorn food crops,

* This is an edited, reformatted and augmented version of Congressional Research Service Publication No. R40529, updated June 27, 2019.

nonfood crops, crop residues, animal manure, and more—as renewable energy sources for liquid fuel use or for power generation. This is partly due to the variety, lack of supply, and dispersed location of noncorn-based biomass feedstock. The technology development status and costs to convert noncorn-based biomass into energy are also viewed by some as obstacles to rapid adoption or deployment.

To aid in understanding the role of biomass as an energy resource, this chapter examines the characterization of biomass in legislation. For over 40 years, the term biomass has been used in legislation enacted by Congress for various programs. Biomass-related legislation has provided financial incentives to develop technologies that use biomass. How biomass is defined influences decisions about the type of biomass that is grown, where it is grown, and potential preferred energy uses, among other things. There have been 14 biomass definitions included in legislation—including tax legislation—since 2004.

Future policy discussions about both energy—particularly legislation involving the Renewable Fuel Standard (RFS) and energy tax incentives—and the environment may prompt further discussion about the definition of biomass. For example, one point of contention regarding the biomass definition and the RFS is whether the term should be defined to include a larger quantity of biomass from federal lands. Some argue that removal of biomass from these lands may lead to ecological harm. Others contend that biomass from federal lands can aid the production of renewable energy to meet certain mandates (e.g., the RFS) and that the removal of biomass can enhance forest restoration or protection from wildfires, insects, or diseases.

Bills have been introduced—in the current Congress and previous Congresses—that would modify the biomass definition, indicating some interest in expanding or refining its use. For example, S. 1614 would expand the renewable biomass definition for the RFS to include more biomass material from federal lands. Bills that would modify the biomass definition were also introduced in the 111th, 112th, and 113th Congresses. This chapter lists biomass definitions enacted by Congress in legislation since 2004, and discusses the similarities and differences among the definitions.

INTRODUCTION

The potential for biomass to meet U.S. renewable energy demands has yet to be fully explored. Nonfood and other types of biomass (e.g., manure) have traditionally been considered by some as waste material and as such have been deposited in landfills, used for animal feed, or applied to crop production lands. However, fuel prices, environmental concerns, and sustainability issues have led policymakers to propose and enact legislation that would encourage conversion of biomass into liquid fuels (e.g., ethanol, biodiesel), electricity, or thermal energy.[1] Since at least 2007, there has been increasing interest in cellulosic biomass (e.g., crop residues, prairie grasses, and woody biomass) because it does not compete directly with crop production for food— although it may compete for land—and because it is located in widely dispersed areas.[2] Classification of biomass as an energy resource has prompted the investigation of its use for purposes other than liquid fuel (e.g., on-site heating and lighting purposes, off-site electricity).

BIOMASS

Biomass is organic matter that can be converted into energy. Common examples of biomass include food crops, crops for energy (e.g., switchgrass or prairie perennials), crop residues (e.g., corn stover), wood waste and byproducts (both mill residues and traditionally noncommercial biomass in the woods), and animal manure. In recent years, the concept of biomass has grown to include such diverse sources as algae, construction debris, municipal solid waste, yard waste, and food waste. Some argue that biomass is a renewable resource that is widely available, may be obtained at minimal cost, and may produce less greenhouse gas than fossil fuels under certain situations. Others contend that biomass has seen limited use as an energy source thus far because it is not readily available as a year-round feedstock, is often located at dispersed sites, can be expensive to

---

[1] For more information on biofuels and biopower, CRS Report R41282, *Agriculture-Based Biofuels: Overview and Emerging Issues*, by Mark A. McMinimy, and CRS Report R41440, *Biopower: Background and Federal Support*, by Kelsi Bracmort.

[2] The interest in cellulosic biomass is partly due to the expanded Renewable Fuel Standard (RFS) in the Energy Independence and Security Act of 2007 (P.L. 110-140), which required that specific volumes of cellulosic biofuel be included in transportation fuel starting in 2010. For more information on the cellulosic biofuel requirement for the RFS, see CRS Report R41106, *The Renewable Fuel Standard (RFS): Cellulosic Biofuels*, by Kelsi Bracmort.

transport, lacks long-term performance data, requires costly technology to convert to useful energy, and might not meet quality specifications to reliably fuel electric generators.

Woody biomass has received special attention because of its widespread availability, but to date has been of limited use for energy production except for wood wastes at sawmills. Wood can be burned directly, usually to produce both heat or steam and electricity (called combined heat and power, or CHP), or digested to produce liquid fuels. Biomass from forests, as opposed to mill wastes, has been of particular interest, because it is widely accepted that many forests have excess biomass (compared to historical levels), often referred to as hazardous fuels.[3] Removing these hazardous fuels from forests could reduce the threat of uncharacteristic catastrophic wildfires, or slow the spread of insect or disease infestation, at least in some ecosystems, while providing a feedstock for energy production.

LEGISLATIVE HISTORY

The term biomass was first used by Congress in the Powerplant and Industrial Fuel Use Act of 1978 (P.L. 95-620), where it was referred to as a type of alternate fuel. However, the term was first defined in the Energy Security Act of 1980 (P.L. 96-294, Title II), as "any organic matter which is available on a renewable basis, including agricultural crops and agricultural wastes and residues, wood and wood wastes and residues, animal wastes, municipal wastes, and aquatic plants." Additionally, the Energy Security Act of 1980 excluded certain types of biomass (i.e., aquatic plants and municipal waste) for certain sections of the bill (e.g., Section 252 Biomass Energy Research and Demonstration Projects).

Three laws contain pertinent biomass definitions: the Food, Conservation, and Energy Act of 2008 (2008 farm bill, P.L. 110-246); the Energy Independence and Security Act of 2007 (EISA, P.L. 110-140); and

[3] See CRS Report R40811, *Wildfire Fuels and Fuel Reduction*, by Katie Hoover.

the Energy Policy Act of 2005 (EPAct05, P.L. 109-58). The term is mentioned several times throughout the three laws, but is not defined for each provision of the laws. In some cases, an individual law has multiple biomass definitions related to various provisions. For example, EISA provides three separate definitions and EPAct05 has six biomass definitions. In contrast, the 2008 farm bill has one definition (Title IX). The tax code contains four additional definitions. In total, 14 biomass definitions have been included in legislation—including tax legislation—since 2004. Table 1 includes definitions from the three laws and from the tax code, and contains additional comments.

The definitions are built into the many provisions and programs that may support research and development, encourage technology transfer, and reduce technology costs for landowners and businesses. Thus, because the various definitions determine which feedstocks can be used under the various programs, the definitions are critical to the research, development, and application of biomass used to produce energy.

Analysis of Biomass Definitions

Of the many biomass definitions, there are two that are commonly referenced for energy production purposes: the definition in the 2008 farm bill and the definition in Title II of EISA (see Table 1). Both laws include an extensive definition of renewable biomass, recognizing that biomass is considered by some to be an infinite feedstock that may be replenished in a short time frame. Both definitions consider crops, crop residues, plants, algae, animal waste, food waste, and yard waste, among other items, as appropriate biomass feedstock. The two definitions differ, however, in where the biomass may be obtained from.

Whether the biomass is grown on federal lands is an important distinction between the two definitions for renewable biomass. The 2008 farm bill includes biomass from federal lands as a biofuel feedstock. In contrast, the renewable biomass definition for the RFS under EISA does not include planted trees, tree residue, or slash and pre-commercial thinnings from federal lands or crops from forested land as biofuel feedstocks.[4] In short, the renewable biomass definition in EISA does not make available the majority of woody biomass on federal lands. There has been some congressional interest in expanding the EISA definition to include more biomass from federal lands to better meet the biofuels usage mandated by the Renewable Fuel Standard (RFS).[5] Legislation has been introduced but not enacted.

EISA expanded the RFS and restricted what qualified as renewable biomass. Advocates of the renewable biomass definition in EISA include groups who favor minimal land disturbance (for ecological reasons as well as to sustain sequestered carbon)[6] and are concerned that incentives to use wood waste might increase land disturbance, especially timber harvesting on federal lands. Opponents of this definition include groups who seek to use materials from federal lands and other forested lands (i.e., not tree plantations) as a source of renewable energy while possibly contributing to long-term, sustainable management of those lands.

Advocates of the renewable biomass definition in the 2008 farm bill include groups who seek to use the potentially substantial volumes of waste, woody biomass from federal lands and other (nonplantation) forest lands (e.g., waste from timber harvests, from pre-commercial thinnings, or from wildfire fuel reduction treatments) as a source of renewable energy. Opponents include groups who seek to preserve forested land and federal land, and who are concerned that incentives for using wood waste would encourage activities that could disturb forest lands, possibly damaging important wildlife habitats and water quality, as well as releasing carbon from forest soils.

---

[4] The Renewable Fuel Standard (RFS) is a provision established by the Energy Policy Act of 2005 requiring gasoline to contain a minimum amount of fuel produced from renewable biomass. For more information on the RFS, see CRS Report R43325, *The Renewable Fuel Standard (RFS): An Overview*, by Kelsi Bracmort.

[5] For example, S. 1614 would allow biomass from federal lands including slash and pre-commercial thinnings derived from ecological restoration activities, biomass materials that are derived within the wildland-urban interface, and biomass materials that are obtained from priority areas for authorized hazardous fuels reduction projects or wildfire threat.

[6] Sequestered carbon is the storage of carbon in plants, soils, oceans, or geologic formations.

POTENTIAL ISSUES FOR BIOMASS FEEDSTOCK DEVELOPMENT

It is not clear whether the biomass definitions in the 2008 farm bill and in EISA constitute a barrier to biomass feedstock development for conversion to energy. Concerns for some landowners and business entities that wish to enter the biomass feedstock market include economic stability, risk/reward ratio, revenue generation, land use designation, and lifecycle greenhouse gas emissions. Additionally, the feedstock development potential of woody biomass varies by region. For example, biomass stock tends to be located on private forest land in the southeastern United States and on federal land in the western United States. Different regions may require different resources to develop a robust biomass feedstock market.

There is mixed support for biomass use, including its feedstock development. Recent agricultural and energy legislation has incorporated provisions and established programs to promote the development and use of biomass as a renewable energy source.[7] However, there also have been efforts in Congress to stall or prevent the use of biomass for energy production, which in turn would impact biomass feedstock development. For instance, there have been repeated attempts to both modify and eliminate the RFS.[8] Further, the 113th Congress enacted a provision that requires that the Department of Defense (DOD) notify Congress at least 30 days prior to the department entering any contract to plan, design, refurbish or construct a biofuel refinery (P.L. 113-291).

Additionally, the success of the provisions and programs that support biomass as a renewable feedstock will be partly determined by landowner participation rates. Participation rates may depend on the definition

---

[7] For more information, see CRS In Focus IF10288, *Overview of Bioenergy Programs in the 2018 Farm Bill*, by Kelsi Bracmort, and CRS Report R40913, *Renewable Energy and Energy Efficiency Incentives: A Summary of Federal Programs*, by Lynn J. Cunningham.

[8] For example, H.R. 104 would reduce the amount of renewable fuel that must be contained in gasoline, among other things. H.R. 1314 (115th Congress) and S. 1584 (114th Congress) would have repealed the RFS. H.R. 4849 (113th Congress) would have eliminated eligibility for corn-starch ethanol in the program and required advanced biofuel eligible for the RFS to have been produced in the United States, among other things.

provided in the legislation, particularly if it authorizes financial and technical support. For example, landowners are eligible to receive financial or technical assistance for biomass feedstock development based on the renewable biomass definition for a specific program.[9]

PROPOSED MODIFICATION OF THE BIOMASS DEFINITION

The various definitions for biomass contained in statute determine what qualifies as biomass and which lands are eligible for biomass removal for the RFS, farm bill energy programs, tax incentives, and more. How biomass is defined in legislation influences decisions on the types of crops grown, where they are grown, and their potential preferred energy uses, among other things. Biomass definitions typically contain three components: agriculture (e.g., crops), forestry (e.g., slash, pre-commercial thinnings), and waste (e.g., food, yard). Multiple biomass definitions can be included in a single piece of legislation to meet the requirements of associated programs or provisions. Environmental groups, private entities aspiring to participate in biomass-to-energy initiatives, and federal agencies that administer biomass-to-energy programs are likely to closely monitor biomass definitions proposed during future energy, environmental, and farm bill debates in Congress.

Thus far, biomass debate and action in the 116th Congress has been minimal. Debates about the biomass definition have occurred in previous Congresses (e.g., 111th Congress).[10]

---

9 For instance, the Biomass Crop Assistance Program (BCAP), established in the 2008 farm bill, is intended to support the establishment and production of eligible crops for conversion to bioenergy. For more information on BCAP, see CRS Report R41296, *Biomass Crop Assistance Program (BCAP): Status and Issues*, by Mark A. McMinimy.

10 For example, bills were introduced in the 113th Congress that would have modified the biomass definition (H.R. 4426, H.R. 4956, H.R. 3084, S. 1267). Additionally, in the 112th Congress, bills were introduced to modify the biomass definition including S. 559, S. 781, H.R. 1861, and H.R. 1920. Significant attention was focused on the proposed biomass definitions contained in multiple legislative proposals put forth by the 111th Congress, including the American Clean Energy and Security Act of 2009 (ACES; H.R. 2454), the

Table 1. Biomass definitions contained in legislation enacted since 2004

No.[0]	Public Law/ Tax Code	Definition	Comments
1	P.L. 110-246 Food, Conservation, and Energy Act of 2008 (2008 farm bill) Title IX Section 9001(12)	The term *renewable biomass* means— '(A) materials, pre-commercial thinnings, or invasive species from National Forest System land and public lands (as defined in section 103 of the Federal Land Policy and Management Act of 1976 (43 U.S.C. 1702)) that—'(i) are byproducts of preventive treatments that are removed—'(I) to reduce hazardous fuels; '(II) to reduce or contain disease or insect infestation; or '(III) to restore ecosystem health; '(ii) would not otherwise be used for higher-value products; and '(iii) are harvested in accordance with—'(I) applicable law and land management plans; and '(II) the requirements for— '(aa) old-growth maintenance, restoration, and management direction of paragraphs (2), (3), and (4) of subsection (e) of section 102 of the Healthy Forests Restoration Act of 2003 (16 U.S.C. 6512); and '(bb) large-tree retention of subsection (f) of that section; or '(B) any organic matter that is available on a renewable or recurring basis from non-Federal land or land belonging to an Indian or Indian tribe that is held in trust by the United States or subject to a restriction against alienation imposed by the United States,	Definition applies to the following sections in the bill: Biorefinery Assistance Program (§9003), Repowering Assistance (§9004), Biomass Research and Development Initiative (§9008), Biomass Crop Assistance Program (§9011), Forest Biomass for Energy (§9012), and Community Wood Energy Program (§9013). In contrast to the definition of *renewable biomass* for the Renewable Fuel Standard (RFS) in P.L. 110-140 discussed below, this definition more broadly allows biomass from federal lands as a biofuel feedstock. The definition *includes* materials, pre-commercial thinnings, or invasive species from National Forest System land and public (BLM) lands; organic matter available on a renewable or recurring basis from non-federal or Indian land; renewable plant material (e.g., feed grains, other agricultural commodities, other plants and trees, and algae) and waste material (e.g., crop residue and other vegetative waste material, such as wood waste and wood residues), as well as animal waste and byproducts (including fats, oils, greases, and manure), food waste, and yard waste.

American Clean Energy Leadership Act of 2009 (ACELA; S. 1462), the Clean Energy Jobs and American Power Act (S. 1733), and the American Power Act (discussion draft). Discussion regarding the definition of biomass during the 111th Congress tended to center on the type of forestry products considered as an eligible biomass source and the lands (e.g., federal, forested) where biomass removal can occur. The eligibility of forest products may have been a contentious aspect of the biomass definition primarily because of differing viewpoints on the sustainability of woody biomass supplies. Some voice disapproval about forest lands being eligible for biomass removal generally because it is uncertain whether forestry products can be removed and transported to an energy conversion facility with minimal environmental impact, and whether such removals damage forest health. Others contend that inclusion of biomass removal from federal and forested lands is necessary to meet specific biofuel mandates established in the RFS.

Table 1. (Continued)

No.[0]	Public Law/ Tax Code	Definition	Comments
		including—'(i) renewable plant material, including—'(I) feed grains; '(II) other agricultural commodities; '(III) other plants and trees; and '(IV) algae; and '(ii) waste material, including—'(I) crop residue; '(II) other vegetative waste material (including wood waste and wood residues); '(III) animal waste and byproducts (including fats, oils, greases, and manure); and '(IV) food waste and yard waste.	This definition does *not* mention biomass from wildfire fuel treatments in the immediate vicinity of buildings, as does the RFS *renewable biomass* definition. (See definition 2.) No limits on private-sector participation evident.
2	P.L. 110-140 Energy Independence and Security Act of 2007 (EISA) Title II Section 201(1)(I)	The term *renewable biomass* means each of the following: "(i) Planted crops and crop residue harvested from agricultural land cleared or cultivated at any time prior to the enactment of this sentence that is either actively managed or fallow, and nonforested. "(ii) Planted trees and tree residue from actively managed tree plantations on non-federal land cleared at any time prior to enactment of this sentence, including land belonging to an Indian tribe or an Indian individual, that is held in trust by the United States or subject to a restriction against alienation imposed by the United States. "(iii) Animal waste material and animal byproducts. "(iv) Slash and pre-commercial thinnings that are from non-federal forestlands, including forestlands belonging to an Indian tribe or an Indian individual, that are held in trust by the United States or subject to a restriction against alienation imposed by the United States, but not forests or forestlands that are ecological communities with a global or State ranking of critically imperiled, imperiled, or rare pursuant to a State Natural Heritage Program, old growth forest, or late successional forest. "(v) Biomass obtained from the	This provision defines renewable biomass for the Renewable Fuel Standard (RFS). The definition does not include, as biofuel feedstocks, planted trees, tree residue, or slash and pre-commercial thinnings removed from federal lands or crops from forested land (e.g., timber harvests).The definition *includes* biomass from: • slash and pre-commercial thinnings from non-federal forestlands, including Indian forestlands, but not from forestlands that are ecological communities within a global or state ranking of critically imperiled, imperiled, or rare pursuant to a State Natural Heritage Program, and not from old-growth or late successional forests; • planted trees and tree residue from actively managed tree plantations on non-federal land; • biomass obtained from the immediate vicinity of buildings, public infrastructure, and areas regularly occupied by people that are at risk from wildfire; and • other activities, including planted crops and crop residue from nonforested agricultural land that is either

No.[0]	Public Law/ Tax Code	Definition	Comments
		immediate vicinity of buildings and other areas regularly occupied by people, or of public infrastructure, at risk from wildfire. "(vi) Algae. "(vii) Separated yard waste or food waste, including recycled cooking and trap grease.	actively managed or fallow; animal waste material and byproducts; separated yard waste or food waste (including recycled cooking and trap grease); and algae. Some biomass material may come from non-federal forest lands.
3	P.L. 110-140 Energy Independence and Security Act of 2007 (EISA) Title XII Section 1201	The term *biomass*—'(aa) means any organic material that is available on a renewable or recurring basis, including—'(AA) agricultural crops; '(BB) trees grown for energy production; '(CC) wood waste and wood residues; '(DD) plants (including aquatic plants and grasses); '(EE) residues; '(FF) fibers; '(GG) animal wastes and other waste materials; and '(HH) fats, oils, and greases (including recycled fats, oils, and greases); and '(bb) does not include—'(AA) paper that is commonly recycled; or '(BB) unsegregated solid waste.	Definition associated with this section in the bill: Express Loans for Renewable Energy and Energy Efficiency (§1201). The definition *excludes* paper that is commonly recycled and unsegregated solid waste. The definition *includes* any organic material available on a renewable or recurring basis, including agricultural crops, trees grown for energy production, wood waste and wood residues, plants (including aquatic plants and grasses), residues, fibers, animal wastes and other waste materials, and fats, oils, and greases (including recycled fats, oils, and greases). The definition is less restrictive than the *renewable biomass* definition listed under EISA, Title II, Section 201(1)(I) for the RFS. The definition is similar to the biomass definition under the EPAct05 Renewable Energy Security Provision (§206). Does not specify "actively managed" crops and trees as a criterion as mentioned in definition 2.
4	P.L. 110-140 Energy Independence and Security Act of 2007 (EISA) Title XII Section 1203(e)(z)(4)(A)	The term *biomass*—'(i) means any organic material that is available on a renewable or recurring basis, including—'(I) agricultural crops; '(II) trees grown for energy production; '(III) wood waste and wood residues; '(IV) plants (including aquatic plants and grasses); '(V) residues; '(VI) fibers; '(VII) animal wastes and other waste materials; and '(VIII) fats, oils, and greases (including recycled fats, oils, and greases);	Definition associated with this section in the bill: Small Business Energy Efficiency Program (§1203). Applicable biomass is identical to materials described in definition 3.

Table 1. (Continued)

No.[0]	Public Law/ Tax Code	Definition	Comments
		and '(ii) does not include—'(I) paper that is commonly recycled; or '(II) unsegregated solid waste.	
5	Tax Code 2007 Title 26 Subtitle A Chapter 1 Subchapter A Part IV Subpart D Section 45(c)(2)	The term *closed-loop biomass* means any organic material from a plant which is planted exclusively for purposes of being used at a qualified facility to produce electricity.	Definition associated with this section in the code: Electricity Produced from Certain Renewable Resources (§45). Definition associated with this tax credit: Renewable Electricity, Refined Coal, and Indian Coal Production Credit (IRS Form 8835). Denotes the following as applicable biomass: • any organic material from a plant that is grown exclusively to produce electricity.
6	Tax Code 2007 Title 26 Subtitle A Chapter 1 Subchapter A Part IV Subpart D Section 45(c)(3)	The term *open-loop biomass* means—(i) any agricultural livestock waste nutrients, or (ii) any solid, nonhazardous, cellulosic waste material or any lignin material which is derived from—(I) any of the following forest-related resources: mill and harvesting residues, precommercial thinnings, slash, and brush, (II) solid wood waste materials, including waste pallets, crates, dunnage, manufacturing and construction wood wastes (other than pressure-treated, chemically-treated, or painted wood wastes), and landscape or right-of-way tree trimmings, but not including municipal solid waste, gas derived from the biodegradation of solid waste, or paper which is commonly recycled, or (III) agriculture sources, including orchard tree crops, vineyard, grain, legumes, sugar, and other crop by-products or residues. Such term shall not include closed-loop biomass or biomass burned in conjunction with fossil fuel (cofiring) beyond such fossil fuel required for startup and flame stabilization.	Definition associated with this section in the code: Electricity Produced from Certain Renewable Resources (§45). Definition associated with this tax credit: Renewable Electricity, Refined Coal, and Indian Coal Production Credit (IRS Form 8835). Denotes the following as applicable biomass: • any agricultural livestock waste nutrients. • any solid, nonhazardous, cellulosic waste material or lignin material. Does not include municipal solid waste, gas derived from the biodegradation of solid waste, or paper which is commonly recycled. Does not include closed-loop biomass or biomass burned in conjunction with fossil fuel (cofiring) beyond such fossil fuel required for startup and flame stabilization.

No.[0]	Public Law/ Tax Code	Definition	Comments
7	Tax Code 2007 Title 26 Subtitle A Chapter 1 Subchapter A Part IV Subpart D Section 45k(c)(3)	The term *biomass* means any organic material other than—(A) oil and natural gas (or any product thereof), and (B) coal (including lignite) or any product thereof.	Definition associated with this section in the code: Tax Credit for Producing Fuel from a Nonconventional Source (§45k). Denotes the following as applicable biomass: • any organic material other than oil and natural gas, and coal or any product thereof. Definition does not distinguish between open-loop biomass and closed-loop biomass. Definition appears to be more expansive than definitions provided in P.L. 110-140 and P.L. 110-246.
8	Tax Code 2007 Title 26 Subtitle A Chapter 1 Subchapter A Part IV Subpart E Section 48b(c)(4)	The term *biomass* means any—(i) agricultural or plant waste, (ii) byproduct of wood or paper mill operations, including lignin in spent pulping liquors, and (iii) other products of forestry maintenance. (B) Exclusion: The term "biomass" does not include paper which is commonly recycled.	Definition associated with this section in the code: Qualifying Gasification Project Credit (§48b). Denotes the following as applicable biomass: • any agricultural or plant waste. • wood or paper mill operations byproduct. • other products of forestry maintenance. Does not include paper which is commonly recycled. Does not include closed-loop biomass.
9	P.L. 109-58 Energy Policy Act of 2005 (EPAct05) Title II Section 203(b)(1)	The term *biomass* means any lignin waste material that is segregated from other waste materials and is determined to be nonhazardous by the Administrator of the Environmental Protection Agency and any solid, nonhazardous, cellulosic material that is derived from—(A) any of the following forest-related resources: mill residues, precommercial thinnings, slash, and brush, or nonmerchantable material; (B) solid wood waste materials, including waste pallets, crates, dunnage, manufacturing and construction wood wastes (other than pressure-treated, chemically treated, or painted wood wastes), and landscape or right-of-way	Definition associated with this section in the bill: Federal Government Purchase Requirement for Renewable Energy (§203). Denotes the following as applicable biomass: • any lignin waste material that is segregated from other waste materials. • any solid, nonhazardous, cellulosic material derived from forest-related resources, solid wood waste materials, agriculture wastes, or a plant that is grown exclusively as a fuel for the production of electricity. Does not include municipal solid waste, gas derived from the

Table 1. (Continued)

No.[0]	Public Law/ Tax Code	Definition	Comments
		tree trimmings, but not including municipal solid waste (garbage), gas derived from the biodegradation of solid waste, or paper that is commonly recycled; (C) agriculture wastes, including orchard tree crops, vineyard, grain, legumes, sugar, and other crop by-products or residues, and livestock waste nutrients; or (D) a plant that is grown exclusively as a fuel for the production of electricity.	biodegradation of solid waste, or paper that is commonly recycled. Introduces a concept that will be defined in 2007 as "closed-loop biomass" in the Internal Revenue Code. Does not specify "actively managed" crops and trees as a criterion as mentioned in definition 2. Does not discuss biomass obtained from the immediate vicinity of buildings (see definition 2). Limits on private-sector participation not specified.
10	P.L. 109-58 Energy Policy Act of 2005 (EPAct05) Title II Section 206(a)(6)(B)	The term *biomass* means any organic matter that is available on a renewable or recurring basis, including agricultural crops and trees, wood and wood wastes and residues, plants (including aquatic plants), grasses, residues, fibers, and animal wastes, municipal wastes, and other waste materials	Definition associated with this section in the bill: Renewable Energy Security Provision (§206). Denotes the following as applicable biomass: • any organic matter available on a renewable or recurring basis including agricultural. • crops and trees, wood and wood wastes and residues, plants (including aquatic plants), grasses, residues, fibers, and animal wastes, municipal wastes, and other waste materials. Limits on private-sector participation not specified
11	P.L. 109-58 Energy Policy Act of 2005 (EPAct05) Title II Section 210(a)(1)	The term *biomass* means nonmerchantable materials or precommercial thinnings that are byproducts of preventive treatments, such as trees, wood, brush, thinnings, chips, and slash, that are removed—(A) to reduce hazardous fuels; (B) to reduce or contain disease or insect infestation; or (C) to restore forest health.	Definition associated with this section in the bill: Grants to Improve Commercial Value of Forest Biomass for Electric Energy, Useful Heat, Transportation Fuels and Other Commercial Purposes Program (§210). Denotes the following as applicable biomass: • unmarketable materials or precommercial thinnings that are byproducts of preventive treatments, such as trees, wood, brush, thinnings, chips, and slash. Definition limited to forestry

No.[0]	Public Law/ Tax Code	Definition	Comments
			biomass sources. Limits on private-sector participation not specified.
12	P.L. 109-58 Energy Policy Act of 2005 (EPAct05) Title IX Subtitle C Section 932(a)(1)	The term *biomass* means—(A) any organic material grown for the purpose of being converted to energy; (B) any organic byproduct of agriculture (including wastes from food production and processing) that can be converted into energy; or (C) any waste material that can be converted to energy, is segregated from other waste materials, and is derived from—(i) any of the following forest-related resources: mill residues, precommercial thinnings, slash, brush, or otherwise nonmerchantable material; or (ii) wood waste materials, including waste pallets, crates, dunnage, manufacturing and construction wood wastes (other than pressure-treated, chemically-treated, or painted wood wastes), and landscape or right-of-way tree trimmings, but not including municipal solid waste, gas derived from the biodegradation of municipal solid waste, or paper that is commonly recycled.	Definition associated with this section in the bill: Bioenergy Program (§932). Denotes the following as applicable biomass: • any organic material grown for the purpose of being converted to energy. • any organic byproduct of agriculture. • any waste material that can be converted to energy, is segregated from other waste materials, and is derived from forest-related resources or wood waste materials, and landscape or right-of-way tree trimmings. Does not include municipal solid waste, gas derived from the biodegradation of municipal solid waste, or paper that is commonly recycled. Does not specify "actively managed" crops and trees as a criterion as mentioned in definition 2. Limits on private-sector participation not specified.
13	P.L. 109-58 Energy Policy Act of 2005 (EPAct05) Title XIII Subtitle A Section 1307 Section 48B(c)(4)	The term *biomass* means any—'(i) agricultural or plant waste, '(ii) byproduct of wood or paper mill operations, including lignin in spent pulping liquors, and '(iii) other products of forestry maintenance. '(B) EXCLUSION- The term 'biomass' does not include paper which is commonly recycled.	Definition associated with this section in the bill: Credit for Investment in Clean Coal Facilities (§1307). Denotes the following as applicable biomass: • any agricultural or plant waste. • any byproduct of wood or paper mill operations. • any other products of forestry maintenance. Does not include paper which is commonly recycled. Does not specify "actively managed" crops and trees as a criterion as mentioned in definition 2. Limits on private-sector participation not specified.

Table 1. (Continued)

No.[0]	Public Law/ Tax Code	Definition	Comments
14	P.L. 109-58 Energy Policy Act of 2005 (EPAct05) Title XV Subtitle A Section 1512(r)(4)(B)	The term *renewable biomass* is, as defined in Presidential Executive Order 13134, published in the *Federal Register* on August 16, 1999, any organic matter that is available on a renewable or recurring basis (excluding old-growth timber), including dedicated energy crops and trees, agricultural food and feed crop residues, aquatic plants, animal wastes, wood and wood residues, paper and paper residues, and other vegetative waste materials. Old-growth timber means timber of a forest from the late successional stage of forest development.	Definition associated with this section in the bill: Conversion Assistance for Cellulosic Biomass, Waste-Derived Ethanol, Approved Renewable Fuels Grants Program (§1512). Denotes the following as applicable biomass: • any organic matter that is available on a renewable or recurring basis (excluding old-growth timber) including dedicated energy crops and trees, agricultural food and feed crop residues, aquatic plants, animal wastes, wood and wood residues, paper and paper residues, and other vegetative waste materials. Does not specify "actively managed" crops and trees as a criterion as mentioned in definition 2. Limits on private-sector participation not specified.

Source: P.L. 110-246 (Food, Conservation, and Energy Act of 2008), P.L. 110-140 (Energy Independence and Security Act of 2007), P.L. 109-58 (Energy Policy Act of 2005), P.L. 110-172 (Tax Technical Corrections Act of 2007, P.L. 102-486 (Energy Policy Act of 1992), P.L. 101-58 (Omnibus Budget Reconciliation Act of 1990).

Notes:

[a] Numerical listing of definitions does not reflect any order of importance.

Forthcoming congressional consideration of energy issues, particularly legislation involving the RFS or energy tax incentives, and environmental issues may prompt further discussion about the biomass definition in the 116th Congress.

In: Key Congressional Reports for June 2019 ISBN: 978-1-53616-662-0
Editor: Stephanie A. Betts

Chapter 5

CRITICAL MINERALS AND U.S. PUBLIC POLICY*

Marc Humphries

SUMMARY

President Trump and various U.S. lawmakers have expressed concerns about U.S. reliance on critical mineral imports and potential disruption of supply chains that use critical minerals for various end uses, including defense and electronics applications. Chinese export quotas on a subset of critical minerals referred to as rare earth elements (REEs) and China's 2010 curtailment of REE shipments to Japan heightened U.S. vulnerability concern.

In December 2017, Presidential Executive Order 13817, "A Federal Strategy to Ensure Secure and Reliable Supplies of Critical Minerals," tasked the Department of the Interior to coordinate with other executive branch agencies to publish a list of critical minerals. The Department of the Interior published a final list of 35 critical minerals in May 2018.

The concern among many in Congress has evolved from REEs and REE supply chains to include other minor minerals and metals that are

---

* This is an edited, reformatted and augmented version of Congressional Research Service, Publication No. R45810, dated June 28, 2019.

used in small quantities for a variety of economically significant applications (e.g., laptops, cell phones, electric vehicles, and renewable energy technologies) and national defense applications. Also, as time passed, concerns increased about access to and the reliability of entire supply chains for rare earths and other minerals. Congressional action (e.g., National Defense Authorization Act for FY2014, P.L. 113-66) has led to the acquisition of REEs and other materials for the National Defense Stockpile. In 2017, the United States had no primary production of 22 minerals and was limited to byproduct production of 5 minerals on the critical minerals list. In contrast, the United States is a leading producer of beryllium and helium, and there is some U.S. primary production of 9 other critical minerals. China ranked as the lead global producer of 16 minerals and metals listed as critical. Although there are no single monopoly producers in China, as a nation, China is a dominant or near-monopoly producer of yttrium (99%), gallium (94%), magnesium metal (87%), tungsten (82%), bismuth (80%), and rare earth elements (80%).

The United States is 100% import reliant on 14 minerals on the critical minerals list (aside from a small amount of recycling). These minerals are difficult to substitute inputs into the U.S. economy and national security applications; they include graphite, manganese, niobium, rare earths, and tantalum, among others. The United States is more than 75% import reliant on an additional 10 critical minerals: antimony, barite, bauxite, bismuth, potash, rhenium, tellurium, tin, titanium concentrate, and uranium.

The current goal of U.S. mineral policy is to promote an adequate, stable, and reliable supply of materials for U.S. national security, economic well-being, and industrial production. U.S. mineral policy emphasizes developing domestic supplies of critical materials and encourages the domestic private sector to produce and process those materials. But some raw materials do not exist in economic quantities in the United States, and processing, manufacturing, and other downstream ventures in the United States may not be globally cost competitive. Congress and other decisionmakers have multiple legislative and administration options to weigh in deliberating on whether, and if so how, to address the U.S. role and vulnerabilities related to critical minerals.

INTRODUCTION

President Trump and various U.S. lawmakers have expressed concerns about U.S. reliance on critical mineral imports and the vulnerability to critical mineral disruptions of supply chains for various end uses, including defense and electronics applications. Chinese export quotas on a type of critical minerals referred to as rare earth elements (REEs) and China's curtailment of rare earth shipments to Japan over a maritime dispute in 2010 represented a wakeup call for the United States on China's near-monopoly control over global REE supply.[1] The actions of the Chinese led to record high prices for REEs and, as a result, began to shine a light on the potential supply risks and supply chain vulnerability for rare earths and other raw materials and metals needed for national defense, energy technologies, and the electronics industry, among other end uses.[2] U.S. legislators have introduced and deliberated on bills that would address the potential supply risk and vulnerability with respect to rare earth supply and bills that would promote domestic rare earth mine development.

After 2010, decisionmakers were faced with various policy questions, including is a domestic supply chain necessary to address potential supply risk; and would an RRE alternative supply chain outside China among allies provide reliable and less risky access to RREs? As events unfolded during the 2010s, it became clear that providing an upstream supply outside China was not enough, and that access to and the reliability of

---

[1] There are 17 rare earth elements (REEs): yttrium, scandium, and 15 within the chemical group called lanthanides. The lanthanides consist of the following: lanthanum, cerium, praseodymium, neodymium, promethium, samarium, europium, gadolinium, terbium, dysprosium, holmium, erbium, thulium, ytterbium, and lutetium. Rare earths are moderately abundant in the earth's crust, some even more abundant than copper, lead, gold, and platinum. While some are more abundant than many other minerals, most REEs are not concentrated enough to make them easily exploitable economically. The lanthanides are often broken into two groups: light rare earth elements (LREEs)—lanthanum through europium (atomic numbers 57-63), and the heavier rare earth elements (HREEs)—gadolinium through lutetium (atomic numbers 64-71). Yttrium is typically classified as a heavy element. Currently, the dominant U.S. end uses for rare earth elements are for automobile catalysts and petroleum refining catalysts; use in phosphors in color television and flat panel displays (cell phones, portable DVDs, and laptops); permanent magnets and rechargeable batteries for hybrid and electric vehicles; and numerous medical devices. There also are defense applications such as jet fighter engines, missile guidance systems, antimissile defense, and satellite and communication systems. Permanent magnets containing neodymium, gadolinium, praseodymium, dysprosium, and terbium are used in numerous electrical and electronic components and new-generation generators for wind turbines.

[2] For more information on critical mineral end uses, see Table 5.

entire supply chains for rare earths and other minerals essential for the economy and national security also were vulnerable.

The concern among many in Congress has evolved from rare earths and REE supply chains, to also include other minor minerals or metals that used in small quantities for a variety of economically significant applications.[3] These minor metals are used in relatively small amounts in everyday applications such as laptops, cell phones and electric vehicles, and renewable energy technologies, in addition to national defense applications. In December 2017, the Presidential Executive Order (E.O.) 13817, "A Federal Strategy to Ensure Secure and Reliable Supplies of Critical Minerals," tasked the Department of the Interior (DOI) to coordinate with other executive branch agencies to publish a list of "critical minerals."[4] DOI published a final list of 35 critical minerals in May 2018.[5]

From 2010 to Present

Initially after China's actions in 2010 contributed to prices for the various elements increasing, the focus in Congress was on rare earth supply (e.g., where in the United States new REE production could begin).

---

[3] Minor metals are primarily byproducts of base metals (e.g., copper, iron, nickel, zinc) and typically not traded on exchanges.

[4] The National Science and Technology Council Subcommittee on Critical and Strategic Mineral Supply Chains (CSMSC) and Executive Order (E.O.) 13817 define critical minerals as those that have a supply chain that is vulnerable to disruption, and that serve an essential function in the manufacture of a product, the absence of which would cause significant economic or security consequence. Strategic minerals are defined as a subset of critical minerals and are essential for national security applications. For more on E.O. 13817, see https://www.federalregister.gov/documents/ 2017/12/26/2017-27899/a-federal-strategy-to-ensure-secure-and-reliable-supplies-of-critical-minerals.

[5] 83 *Federal Register* 23295, May 18, 2018. The December 2018 list of critical minerals includes: aluminum, antimony, arsenic, barite, beryllium, bismuth, cesium, chromium, cobalt, fluorspar, gallium, germanium, graphite, hafnium, helium, indium, lithium, magnesium compounds, manganese, niobium, platinum group metals, rare earth elements, potash, rhenium, rubidium, scandium, strontium, tantalum, tellurium, tin, titanium, tungsten, uranium, vanadium, and zirconium. This is not a static list and is subject to change. For more on E.O. 13817, see section on "Development of the Critical Minerals List" in this chapter.

Since 2010, several bills have been introduced that would use a variety of policy options and approaches—from streamlining the permitting framework for rare earth elements and other mining and processing projects on federal land, to the additions of REEs to the National Defense Stockpile.[6] Sections 1411 and 1412 of the National Defense Authorization Act for FY2014 (P.L. 113-66) contained language for Department of Defense to begin studies of rare earth materials and to require purchases of heavy REEs for the national defense stockpile.

In 2010 the sole U.S. rare earth mine located in Mountain Pass, CA, owned by Molycorp, Inc., was dormant. From the mid-1960s through the 1980s, Molycorp's Mountain Pass mine was the world's dominant source of rare earth oxides. However, by 2000, nearly all of the separated rare earth oxides were imported, primarily from China. Because of China's REE oversupply and lower-cost production, as well as a number of environmental (e.g., a pipeline spill carrying contaminated water) and regulatory issues at Mountain Pass, Molycorp, Inc. ceased production at its mine in 2002.

Between 2010 and 2012, there was some optimism but also criticism over Molycorp Inc.'s approach to reopen the only rare earth mine in the United States and establish a vertically integrated operation including oxide separation, production of metal alloys, and permanent magnet production.[7] A few important questions relevant to a vertically integrated approach were raised then as they are now

- How can a fully integrated supply chain be developed domestically?
- Is a domestic supply chain necessary to address potential supply risk?; and

[6] The National Defense Stockpile (50 U.S.C. §98 et seq.) was established in 1939 to retain stocks of strategic and critical materials, thus, reducing dependence on foreign sources during times of national emergencies. The stockpile was set up for national defense purposes only and not to be used as an economic stockpile.

[7] Permanent magnets are important parts for national defense missile systems, wind turbines, and automobiles.

- With China in a near-monopoly position in all aspects of the rare earth supply chain, would an alternative supply chain outside China among allies provide reliable and less risky access to needed rare earth elements?

Another immediate concern focused on the investment and skill level needed to build-out a reliable supply chain outside of China.

In 2012, Molycorp, Inc., reopened its Mountain Pass mine, and the Lynas Corporation, Ltd. began production in Australia which added more REEs to the global mix—albeit most of the production was in light rare earth elements (LREEs), not the heavy rare earth elements (HREEs) are needed for permanent magnets—the fastest growing use for rare earth elements at the time. Permanent magnets are important parts for national defense missile systems, wind turbines, and automobiles. With higher prices came lower demand as some companies began to use less REEs, try substitutes, or diversify their source of raw material supply outside of China. With China's production (including illegal production), there was more supply than demand for many of the REEs and prices declined. As a result of rapidly falling prices and Molycorp's debt, the Mountain Pass mine was not economically sustainable. Molycorp filed for Chapter 11 bankruptcy protection in June 2015. In June 2017, MP Mine Operations LLC (MPMO) purchased the Mountain Pass mine for $20.5 million. MPMO is an American-led consortium of which the Chinese-owned Leshan Shenghe Rare Earth Company has a 10% nonvoting minority share. In 2018, MMPO reportedly restarted production at Mountain Pass. See Table 1 for Molycorp's timeline. In March 2019, the Chinese government announced a reduction in REE production quotas and suggested that the REE produced in China would be sold only in China for its domestic manufacturing activity.[8]

[8] Tom Daly, "China Sets Lower Rare Earth Output Quotas for First Half of 2019," Reuters, March 15, 2019, https://www.reuters.com/article/us-china-rareearths-quotas.

Table 1. Timeline of selected molycorp, inc.,-related activities

Mid-1960s through 1990s	Molycorp's Mountain Pass mine in the 1960s-1980s was the world's dominant source of rare earth oxides. U.S. production began to rapidly decline in the 1990s, as China's lower cost production began to accelerate.
By 2000	Nearly all of the separated rare earth oxides in the United States were imported, primarily from China.
2002	Because of China's oversupply and lower-cost production, and a number of environmental (e.g., a pipeline spill carrying contaminated water) and regulatory issues at Mountain Pass, Molycorp ceased production at its mine. Since then, the United States has lost nearly all of its capacity in the rare earth supply chain, including intellectual capacity.
2008	Under new ownership, Molycorp embarked upon a campaign to change the rare earth position in the United States with its "mine to magnet" (vertical integration) business model.
2011	Molycorp broke ground for a new separation facility at the Mountain Pass mine to facilitate a proprietary oxide separation process that it had designed to use fewer reagents and recycle the wastewater, thus eliminating the need for a disposal pond. (April) Molycorp acquired the Japanese subsidiary Santoku America in Tolleson, AZ, and renamed it Molycorp Metals and Alloys (MMA). This acquisition was part of the firm's strategy to become a vertically integrated company. It produced both neodymium iron-boron (NdFeB) and samarium cobalt (SmCo) alloys used in the production of permanent magnets. Molycorp Metals and Alloys was the sole U.S. producer of the NdFeB alloy. (April) Molycorp purchased a 90.023% majority interest in AS Silmet (renamed Molycorp Silmet), an Estonian-based rare earth element and rare metals processor. (November) Molycorp entered a joint venture with Daido Steel and Mitsubishi Corporation of Japan to manufacture sintered permanent rare earth (NdFeB) magnets in Japan that were sold on the world market.
2012	(June) Molycorp acquired Neo Materials Technology, Inc., a Toronto-based firm (renamed Molycorp Canada) with rare earth processing and permanent magnet powder facilities in China. Molycorp restarted rare earth production.
2015	(June) Molycorp files for Chapter 11 Bankruptcy Protection.
2016	(August) Neo Performance Materials is established as a private company following the restructuring of Molycorp. Molycorp remains a separate entity as owner of Mountain Pass Mine.
2017	Neo Performance Materials completes an initial public offering (IPO) on the Toronto Stock Exchange.
2017	(June) A consortium, MP Mine Operations, LLC (MPMO)—comprised of JHL Capital Group, LLC (aka MP Materials) (65%); QVT Financial LP (25%); and Leshan Shenghe Rare Earth Company (10%)—purchase Mountain Pass Mine for $20.5 million.
2018	(January) According to MPMO, production at Mountain Pass restarted in January of 2018. Production data were not available at the time of this writing.

Sources: CRS using CRS Report R41347, *Rare Earth Elements: The Global Supply Chain*, by Marc Humphries, and articles from http://www.mining.com including "Molycorp Thrown a Lifeline" (August 31, 2016), and "Mountain Pass Sells for $20.5 Million" (June 16, 2017), by Andrew Topf.

As previously noted, the vulnerability concerned expanded from RREs to critical minerals. Assessments using a criticality matrix identified minerals (such as REEs, cobalt, and tantalum, among others) that could face supply restrictions and result in vulnerabilities to the economy and national security.[9] Broad criticality assessments were prepared by the National Research Council, the Department of Energy (DOE), and the Massachusetts Institute of Technology (MIT) early in the recent discussion of mineral supply risk and potential mineral demand from the energy technology sector.[10] Many others, such as Nassar, Du, and Graedel,[11] have weighed in since 2010 on the criticality and supply risk question, providing a variety of models that examine the supply risk and vulnerabilities associated with these minerals. It is beyond the scope of this chapter to evaluate those models.

Congressional Interest

Proposed Congressional findings mentioned in a number of bills introduced since the 111th Congress on critical minerals include:

- Emerging economies are increasing their demand for REEs as they industrialize and modernize;
- A variety of minerals are essential for economic growth and for infrastructure;

---

[9] A criticality matrix is a two-dimensional presentation of a mineral's importance of use and availability. This framework for analysis seeks to emphasize on a vertical axis whether minerals are easily substitutable, identify the impacts of potential supply restrictions, and on the horizontal axis the potential supply risks associated with geology, ecology, technology, economics, and the political environment. Many analysts have used the criticality matrix to rank the criticality of selected minerals.

[10] National Research Council, Minerals, Critical Minerals, and the U.S. Economy, National Academies Press, 2008; U.S. DOE, *Critical Materials Strategy,* December 2011; American Physical Society and The Materials Research Society, *Energy Critical Elements, Securing Materials for Emerging Technologies*, 2011.

[11] N.T. Nassar, Xiaoyue Du, and T.E. Graedel, "Criticality of the Rare Earth Elements," *Journal of Industrial Ecology*, v. 19, no. 6, 2015.

- The United States has vast mineral resources but at the same time is becoming more dependent on imports;
- Mineral exploration dollars in the United States are approximately 7% of the world total (compared to 19% in the early 90s);
- Heavy rare earth elements are critical to national defense;
- China has near-monopoly control over the rare earth value chain, and there has been a transfer of technology from U.S. firms and others to China in order to gain access to rare earths and downstream materials;
- Thorium regulations are a barrier to rare earth development in the United States;
- A sense of Congress that China could disrupt REE and other critical mineral supplies to the United States;
- It is important to develop the domestic industrial base for the production of strategic and critical minerals; and
- The United States must accept some risk in the form of aiding domestic investment opportunities.

The Senate Committee on Energy and Natural Resources held a hearing on S. 1317, the American Mineral Security Act, on May 14, 2019, "Examining the Path to Achieving Mineral Security."[12] Two congressional hearings were held on critical minerals in the 115th Congress: one on December 12, 2017, by the House Natural Resources Subcommittee on Energy and Mineral Resources on "Examining Consequences of America's Dependence on Foreign Minerals," and a second on July 17, 2018, by the Senate Committee on Energy and Natural Resources to examine the final list of critical minerals.[13]

---

[12] For hearing details, see https://www.energy.senate.gov/public/index.cfm/hearings-and-business-meetings?ID= 559FE490-CA2A-4A56-9C30-F5ED5B0C7780.

[13] For hearing details, see House Natural Resources Subcommittee on Energy and Mineral Resources hearing, https://docs.house.gov/committee/calendar/byevent.aspx?eventID= 106736 (December 21, 2017); and Senate Energy and Natural Resources hearing, https://www.energy.senate.gov/public/index.cfm/2018/7/full-committee-hearing-to-examine-the-department-of-the-interior-s-final-list-of-critical-minerals (July 17, 2018).

Public resource and minerals policy options are among the options for creating reliable supply chains of these minerals and metals. The Administration and many in Congress have combined concerns over import dependence and developing domestic supply into a number of policy proposals that would aim to streamline the permitting process for domestic critical mineral production and possibly open more public lands to mineral exploration. A 2017 U.S. Geological Survey (USGS) report, *Critical Mineral Resources of the United States*, presents its mineral assessments of 23 critical minerals for the nation as a whole, but does not break out what might be available on federal lands, where many of the legislative proposals are directed.[14] Others in Congress want to be sure that if a more efficient permitting process is put in place, all the mechanisms for environmental protection and public input are left intact, if not enhanced.[15]

The Scope of This Chapter

This chapter examines the process by which the critical minerals list was drafted, why these minerals are being classified as critical, where production is taking place, and countries holding the largest reserves of critical minerals. There is a brief review of materials required for lithium-ion batteries and solar and wind energy systems, and a discussion of supply chains for rare earth elements and tantalum. This chapter also presents the statutory and regulatory framework for domestic mineral production, legislative proposals, and congressional and executive branch initiatives (and actions), as well as an overview of U.S. critical mineral policy.

[14] USGS, Critical Mineral Resources of the United States—Economic and Environmental Geology and Prospects for Future Supply, Professional Paper 1802, 2017, http://doi.org/10.3133/pp1802.

[15] U.S. House of Representatives, Committee on Natural Resources, H.Rept. 112-583, *Report Together with Dissenting Views on H.R. 4402: National Strategic and Critical Minerals Production Act of 2012.*

There are a number of policy issues related to U.S. critical minerals, such as trade policy (particularly with China) and conflict minerals,[16] just to name two. Treatment of these issues is beyond the scope of this chapter.

BRIEF HISTORY OF U.S. CRITICAL MINERALS AND MATERIALS POLICY

Minerals for national security have long been a concern in the United States. For example, there were concerns over shortages of lead for bullets during the early 1800s. There were material shortages during WWII and the Korean War that contributed to the formation of the National Defense Stockpile. The current stockpile of strategic and critical minerals and materials was developed to address national emergencies related to national security and defense issues; it was not established as an economic stockpile.

In 1939, after Germany invaded Poland, the Strategic Materials Act of 1939 (50 U.S.C. §98, P.L. 76-117) provided the authority for the United States to establish a strategic materials stockpile. Then in 1946, the Strategic and Critical Materials Stockpiling Act was enacted so that the United States would be prepared for national military emergencies and to prevent material shortages. The 1946 Act (P.L. 79-520) set a target of $2.1 billion of materials to be spent for the stockpile.[17] Congress increased funding for supplying the stockpile to $4 billion over four years (1950-1953). The Defense Production Act of 1950 (50 U.S.C. §4501, P.L.81-774) added $8.4 billion to expand supplies of strategic and critical materials.[18]

---

[16] Conflict minerals are defined as ores that when sold or traded have played key roles in helping to fuel conflict and extensive human rights abuses in far eastern Democratic Republic of the Congo (DRC). The main conflict minerals are tantalum, tin, tungsten, and gold (also known as "3TGs").

[17] The stockpile consisted of many base metals and minerals such as copper, lead, zinc, aluminum, chromite, tin, and quartz among others.

[18] CRS Report 95-5, *The National Defense Stockpile: A Historical Perspective*, by Alfred R. Greenwood, December 14, 1994, p. 2 (out of print; available to congressional clients upon request).

In 1951, President Truman formed the Materials Policy Commission (also known as the Paley Commission) which recommended a stockpile for strategic materials and the use of lower cost foreign sources of supply. President Eisenhower established long term stockpile goals during a national emergency as a way to prevent the shortages that occurred during World War II and the Korean War.

The initial time frame for the duration of the emergency the stockpile was intended to cover was three years, but later reduced to one year. However, with the passage of the 1979 Strategic and Critical Minerals Stockpiling Revision Act (P.L. 96-41), a three-year military contingency was reestablished as a criterion for stockpile goals. Funding for the stockpile was subsequently increased to $20 billion.

During the Cold-War era, the National Defense Stockpile (NDS) had an inventory of large quantities of strategic and critical materials. In the early 1990s, after the Cold War with the Soviet Union, the U.S. Congress supported an upgrade and modernization of the strategic materials stockpile. By FY1993, the National Defense Authorization Act (NDAA) for Fiscal Year 1993 (P.L. 102-484) authorized a major sell-off of 44 obsolete and excess materials in the stockpile such as aluminum metal, ferrochromium, ferromanganese, cobalt, nickel, silver, tin, and zinc.[19] The majority of these materials were sold to the private sector. Proceeds of these sales were transferred to other federal or Department of Defense (DOD) programs.

The Modern Day Stockpile

In 1988, the Secretary of Defense delegated the management of the stockpile to the Undersecretary of Defense for Acquisition, Technology, and Logistics and operational activities of the NDS to the Director of the Defense Logistics Agency (DLA). Among other duties, the DLA manages the day-to-day operations of the stockpile program.

---

[19] GAO, NSIAD-93-60, p. 12. Also, see pp. 37-38 for list of proposed disposals.

The current stockpile contains 37 materials valued at $1.152 billion.[20] Much of the materials are processed metals or other downstream products such as, columbium (niobium) metal ingots, germanium metal, tantalum metal, metal scrap, beryllium rods, quartz crystals, and titanium metal.

Congressional action starting in 2014 led to the acquisition of REEs and other materials for the NDS. The DLA is acquiring six materials based on the NDAA for FY2014: Ferro-niobium; dysprosium metal; yttrium oxide; cadmium-zinc-telluride substrates; lithium-ion precursors; and triamino-trinitrobezene.[21]

In FY2016, the DLA made progress on its FY2014 goals for high-purity yttrium and dysprosium metal. The NDS initiated a program to develop economical methods to recycle REEs from scrap and waste. The goal was to investigate technologies to determine whether recycling is feasible in the United States.[22] Work on this project goal is ongoing.

In addition to acquisitions and upgrades, Congress approved a DOD proposal to sell materials determined to be in excess of program needs as part of the FY2017 NDAA (P.L. 114-328).

Initiatives and Actions on Critical Minerals

Development of the Critical Minerals List

E.O. 13817, "A Federal Strategy to Ensure Secure and Reliable Supplies of Critical Minerals," published on December 20, 2017, tasked the Department of the Interior (DOI) to coordinate with other executive branch agencies in establishing a draft list of critical minerals published in the *Federal Register* 60 days from the initial order. On December 17,

---

[20] U.S. Department of Defense, Office of the Under Secretary of Defense for Acquisition, Technology, and Logistics, *Strategic and Critical Materials Operations Report to Congress*, January 2017, pp. 7-8.

[21] Ibid, p. 9.

[22] Ibid, p. 6.

2017, the Secretary of the Interior issued Secretarial Order (No. 3359, "Critical Mineral Independence and Security") directing the U.S. Geological Survey (USGS) and Bureau of Land Management (BLM) to develop the list.[23] DOI agencies, with cooperation from others (e.g., DOD, DOE, and members of the National Science and Technology Council Subcommittee on Critical and Strategic Mineral Supply Chains [CSMSC]), developed using specific criteria an unranked list of 35 minerals. The Secretary of the Interior issued the final list of critical minerals in May 2018.[24]

The USGS used the critical mineral early warning methodology developed by the CSMSC as its starting point for the draft list.[25] One of the metrics used was the Herfindahl-Hirschman Index which measures the concentration of production by country or company. Another metric used was the Worldwide Governance Index, which was used to ascertain the political volatility of a country and is based on six indicators.[26] The early warning methodology is a two-stage process. The first stage uses the geometric mean of three indicators to determine if the mineral is potentially critical: supply risk (production concentration), production growth (change in market size and geological resources), and market dynamics (price changes). The second stage uses the results of the first stage to determine which of the potentially critical minerals require an in-depth analysis.

In developing the list, the USGS also relied on its net import reliance data;[27] its Professional Paper 1802, (referenced in footnote 14 of this chapter); NDAA FY2018 (P.L. 115-91) from DOD; U.S. Energy Information Administration (EIA) data on uranium; and the input of several subject matter experts. The USGS established a threshold above

[23] Secretary's Order 3359, "Critical Mineral Independence and Security" (December 21, 2017), https://www.doi.gov/ sites/doi.gov/files/uploads/so_criticalminerals.pdf.

[24] 83 *Federal Register* 23295, May 18, 2018.

[25] Draft Critical Mineral List—Summary of Methodology and Background Information—U.S. Geological Survey Technical Input Document in Response to Secretarial Order No. 3359, Open File Report 2018-102, DOI/USGS.

[26] Those six indicators are accountability, political stability and absence of violence, government effectiveness, regulatory quality, rule of law, and control of corruption.

[27] "Net import reliance" refers to the percentage of a mineral commodity used by the United States that must be imported from another country.

which the minerals were deemed to be critical. Some minerals below the threshold that had critical applications were also included on the list. The USGS used a supply chain analysis to include some metals, such as aluminum, because the United States is 100% import reliant on bauxite, the primary source mineral for aluminum production.

The unranked list of 35 minerals does not indicate the levels of criticality for some versus others. This is of note because some earlier studies had shown that the supplies of platinum group metals, REEs, niobium, and manganese are potentially far more vulnerable than lithium, titanium, and vanadium.[28] Further, the REEs are not broken out by element. Some of the heavy rare earth elements have been shown to be more critical and vulnerable to supply shortages than some of the lighter elements.

Other Federal Critical Minerals Actions

In addition to developing a critical minerals list, Congress and various executive branch entities have invested in other actions related to critical minerals. Investment in research and development (R&D) is considered by many experts (e.g., DOE, MIT, and elsewhere)[29] to play a critical role in the support for and development of new technologies that would address three primary areas: greater efficiencies in materials use; substitutes or alternatives for critical minerals; and recycling of critical minerals. Below is a summary of selected current federal R&D, and information and analysis activities on critical minerals at federal agencies.

[28] National Research Council, *Minerals, Critical Minerals, and the U.S. Economy*, National Academies Press, 2008. The list is fluid; mineral with low measure of criticality today may have a higher measure of criticality in months or a few years depending on many variables.

[29] See footnote 10 of this chapter for reports by DOE, MIT, and the National Research Council weighing in on critical minerals.

Department of Energy[30]

Critical Materials Hub

DOE's FY2019 budget request included funding for R&D on rare earth and other critical materials. DOE's "Critical Materials Hub" is conducting R&D on a number of critical material challenges, including "end of life" recycling to help mitigate any possible supply chain disruptions of REEs. Funding for the program was at $25 million, each year, for the past three fiscal years (FY2017-FY2019), as FY2019 is the third year of its second five-year research phase.[31] Congress approved this level of support despite the Trump Administration's proposal to eliminate the program in FY2019 and FY2020. The Critical Materials Hub is funded under the Advanced Manufacturing R&D Consortia within DOE's Energy Efficiency and Renewable Energy Program.

REEs from Coal

Additionally, in FY2019 DOE proposed to launch its Critical Materials Initiative within the Fossil Energy R&D program under the Advanced Coal Energy Systems program to examine new technologies to recover REEs from coal and coal byproducts. Congress had appropriated funding for this project under the National Energy Technology Lab (NETL) R&D program during the Obama Administration, despite no request for funding. For FY2019, the Trump Administration requested $30 million in funding for the Critical Materials Initiative; Congress elected to support the initiative at $18 million.

Critical Minerals Report

In December 2010 and December 2011, DOE issued *Critical Materials Strategy* reports. These reports examine and provide demand forecasts for rare earths and other elements required for numerous energy and electronic

---

[30] Budget information on DOE programs obtained from DOE Budget Highlights, FY2010-FY2019 Congressional Budget Request.

[31] The first phase, funded at about $125 million, ran from FY2012 to FY2016.

applications.[32] An update on this research is forthcoming, according to DOE.[33]

Department of the Interior

The National Minerals Information Center housed within the USGS provides an annual summary of critical mineral activity in its Mineral Commodities Summaries report and Minerals Yearbook.[34] The USGS also provides mineral resource assessments and has in 2017 published a study on 23 mineral commodities, all of which have been listed as critical by the Administration.[35] In 2010, the USGS released a report on the rare earth potential in the United States.[36] A 2017 collaboration between the USGS and the State of Alaska issued a report on critical and precious minerals in Alaska[37] and conducted a geospatial analysis identifying critical mineral potential in Alaska.[38] The results of the analysis provided new information on areas of Alaska that might contain deposits of critical minerals.

Department of Defense

In a DOD-led assessment of the U.S. manufacturing and defense industrial base and supply chain resiliency, there are sections on critical minerals and impacts on national security.[39] The DOD continues to fulfill

---

[32] U.S. Department of Energy, *Critical Materials Report*, December 2011.

[33] Personal communication with Diana Bauer, Director of Energy Systems Analysis and Integration, Office of Energy Policy and Systems Analysis, May 16, 2019.

[34] Department of the Interior, USGS, Mineral Commodity Summaries, https://minerals.usgs.gov/minerals/pubs/mcs/; USGS, *Minerals Yearbook*, https://minerals.usgs.gov/minerals/pubs/myb.html.

[35] U.S. Department of the Interior, USGS, *Critical Mineral Resources of the United States—Economic and Environmental Geology and Prospects for Future Supply*, Professional Paper 1802, 2017.

[36] U.S. Department of the Interior, USGS, *The Principal Rare Earth Elements Deposits of the United States—A Summary of Deposits and a Global Perspective*. USGS Scientific Investigations Report 2010-5220.

[37] U.S. Department of the Interior, USGS, *Geospatial Analysis Identifies Critical Mineral-Resource Potential in Alaska*, fact sheet, March 2017.

[38] USGS, Geospatial Analysis Identifies Critical Mineral-Resource Potential in Alaska, Fact Sheet 2017-3012, March 2017.

[39] U.S. Department of Defense, *Assessing and Strengthening the Manufacturing and Defense Industrial Base and Supply Chain Resiliency of the United States*, Report to Donald J. Trump by the Interagency Task Force in Fulfillment of Executive Order 13806, September 2018.

its stockpile goals for various critical materials and has funded small R&D projects related to rare earths.[40]

In 2009, the Office of Industrial Policy reviewed the rare earth mineral supply chain. The Office of the Secretary of Defense reviewed its National Defense Stockpile and issued a report titled: *Reconfiguration of the National Defense Stockpile Report to Congress*.[41]

As part of the Ike Skelton National Defense Authorization Act for FY2011 (Section 843 of P.L. 111-383), the DOD was required by Congress to prepare an "Assessment and Plan for Critical Rare Earth Materials in Defense Applications" and report to a number of congressional committees by July 6, 2011.[42] A DOD assessment and congressional appropriations supported new stockpile goals for HREEs.

In an April 2012 interview with *Bloomberg News*, the DOD head of industrial policy stated that DOD uses less than 5% of the rare earths used in the United States, and that DOD was closely monitoring the rare earth materials market for any projected shortfalls or failures to meet mission requirements.[43]

White House Office of Science and Technology Policy

In 2010, the White House Office of Science and Technology Policy (OSTP) formed an Interagency Working Group on Critical and Strategic Minerals Supply Chains.[44] The group's focus is to establish critical mineral

---

[40] For stockpile goals, see footnote 20 of this chapter. For small grant programs, see ASM International, "Army Research Lab project to develop U.S. supply chain for rare earth elements," June 10, 2014, http://www.asminternational.org/ home, and Texas Rare Earth Resources, "U.S. Defense Logistics Agency Awards Texas Rare Earth Resources Strategic Materials Research Contract," September 25, 2015, http://www.marketwired.com/press-release.

[41] Department of Defense, Reconfiguration of the National Defense Stockpile Report to Congress, April 2009. https://www.dla.mil/Portals/104/Documents/StrategicMaterials/Reports/Operations%20Report/FY16%20Operations%20Report_FINAL_Website%20Version.pdf.

[42] Letter from the Congress of the United States, directed to The Honorable Leon E. Panetta, U.S. Department of Defense, August 5, 2011.

[43] Ratnam, Gopal, "Rare Earth Shortage Would Spur Pentagon to Action," *Bloomberg News*, April 9, 2012, http://www.bloomberg.com/news/2012-04-09/rare-earths-shortage-would-spur-pentagon-to-action.html.

[44] The group's participants include representatives from the Department of Energy, the Department of Defense, the Department of the Interior, the Department of Commerce, the

prioritization and to serve as an early warning mechanism for shortfalls, to establish federal R&D priorities, to review domestic and global policies related to critical and strategic minerals (e.g., stockpiling, recycling, trade, etc.), and to ensure the transparency of information.

The White House National Science and Technology Council Subcommittee on Critical and Strategic Mineral Supply Chains produced a report describing a screening methodology for assessing critical minerals.[45] The "early warning screening" approach for material supply problems was first included as a U.S. policy goal in the National Materials and Minerals Policy, Research and Development Act of 1980 (30 U.S.C. §1601) (P.L. 96-479).[46]

SUPPLY: CRITICAL MINERALS PRODUCTION AND RESOURCES

Production/Supply

According to the 2019 USGS Mineral Commodity Summaries report,[47] China ranked as the number one producer of 16 minerals and metals listed as critical. While there are no single monopoly producers in China, as a nation China is a near-monopoly producer of yttrium (99%), gallium (94%), magnesium metal (87%), tungsten (82%), bismuth (80%), and rare earth elements (80%). China also produces roughly 60% or more of the

---

Environmental Protection Agency, the Department of State, the Department of Justice, and the Office of the U.S. Trade Representative.

[45] National Science and Technology Council, Committee on Environment Natural Resources, and Sustainability, Subcommittee on Critical and Strategic Mineral Supply Chains, *Assessment of Critical Minerals: Screening Methodology and Initial Applications*, March 2016, https://www.whitehouse.gov/sites/whitehouse.gov/files/images/CSMSC%20Assessment%20of%20Critical%20Minerals%20Report%202016-03-16%20FINAL.pdf.

[46] U.S. Governmental Accountability Office, *Advanced Technologies: Strengthened Federal Approach Needed to Help Identify and Mitigate Supply Risks for Critical Raw Materials*, GAO-16-699, September 2016, p. 9.

[47] USGS, *Mineral Commodity Summaries 2019*, February 2019, https://minerals.usgs.gov/minerals/pubs/mcs/2019/ mcs2019.pdf. The annual report used 2017 data; 2018 is estimated.

world's graphite, germanium, tellurium, and fluorspar. In 2017, the United States had no primary production of 22 minerals and byproduct production of five minerals on the critical minerals list. There is some U.S. primary production of nine minerals, and the United States is a leading producer of beryllium and helium (see Table 2, Figure 1).

China had gains in production that far outpaced the rest of the world. By 2003, China had already dominated in the production of graphite, indium, magnesium compounds, magnesium metal, REEs, tungsten, vanadium, and yttrium; it solidified its number one producing status of these minerals about a decade later. Chinese producers are seeking not only to expand their production capacity at home but to continue to negotiate long-term supply agreements or create equity partnerships around the world, particularly in Africa (cobalt and tantalum), Australia (lithium), and South America (lithium).[48]

The dominant producing region for chromium, manganese, platinum group metals, tantalum, and cobalt is southern Africa. Brazil produces 88% of the world's niobium, and Australia accounts for 58% of the world's lithium production, according to USGS data. According to USGS data, critical minerals dominated by a single producing country include: niobium from Brazil, cobalt from the Democratic Republic of the Congo (DRC), platinum group metals from South Africa, REEs (including yttrium), and tungsten from China.

Production of Minerals and Mineral Resource Potential on Federal Land

Current mineral production information on federal land is not available from the DOI. The Government Accountability Office (GAO) noted in a 2008, report that the DOI does not have the authority to collect information from mine operators on the amount of minerals produced or the amount of

---

[48] Elizabeth C. Economy and Michael Levi, *By All Means Necessary, How China's Resource Quest Is Changing the World*, Council of Foreign Relations, 2014; The Society for Mining, Metallurgy, and Exploration (SMME), "Annual Mining Review," *Mining Engineering*, vol. 60, no. 5, May 2018, p. 47, http://www.miningengineeringmagazine.com.

mineral reserves on public lands, and there is no requirement for operators to report production information to the federal government.[49]

However, previous DOI[50] and GAO[51] reports completed in the early 1990s reported that gold, copper, silver, molybdenum, and lead were the five dominant minerals produced on federal lands under the General Mining Law of 1872 (30 U.S.C. §§21-54). Currently, the vast majority of mining activity on federal lands is for gold in Nevada, based on past DOI information. The DOI report also showed that federal lands mineral production represented about 6% of the value of all minerals produced in the United States. There is uncertainty over how much production of minerals occur on federal lands. Most minerals listed as critical are locatable on U.S. federal lands under the General Mining Law of 1872; comprehensive information on which minerals are located and produced on federal land remains incomplete. An unanswered question is the extent that critical mineral resource potential exists on federal land. Until more is known through mineral resource assessments of federal land, it will be hard to determine the impact of opening federal land to development that is now withdrawn from mineral development.

Some mining advocates support developing domestic supply chains in critical minerals. Other stakeholders support a diversified portfolio of reliable suppliers, particularly if foreign sources are more economic or if domestic production (or manufacturing) is uneconomic, not technically feasible, or environmentally unacceptable.

Byproduct Supply

There are six critical minerals that are classified as byproducts: indium, tellurium, gallium, germanium, cobalt, and rhenium.[52] There are important

---

[49] GAO, *Hardrock Mining: Information on State Royalties and Trends in Mineral Imports and Exports*, GAO-08-849R, July 21, 2008.

[50] U.S. Department of the Interior, Task Force on Mining Royalties, *Economic Implications of a Royalty System for Hardrock Minerals*, August 16, 1993.

[51] GAO, *Mineral Resources: Value of Hardrock Minerals Extracted From and Remaining on Federal Lands*, GAO/rced-92-192, August 1992.

[52] A byproduct is a secondary or additional product associated with production of the main/primary ore and is limited by the output of the main/primary product. Co-products

differences between main product and byproduct supply. Byproduct supply is limited by the output of the main product. For example, the amount of indium recoverable in zinc cannot be more than the quantity of indium in the zinc ore. As production of the main product continues, the byproduct supply may be constrained because a higher price of the byproduct does not increase its supply in the immediate term. Even in the long run, the amount of byproduct that can be economically extracted from the ore is limited. That is, byproduct supply is relatively inelastic (i.e., not particularly responsive to price increases of the byproduct). For byproducts, it is the price of the main product, not the byproduct that stimulates efforts to increase supply. But a high enough byproduct price may encourage new technologies that allow for greater byproduct recovery from the main product. There may be occasions when the main product supply contains more byproduct than is needed to meet demand. If this were the case, byproduct processing facilities would need to be expanded so that byproduct processing capacity would not be a limiting factor in byproduct supply.

Another important difference between byproduct and main product is that only costs associated with byproduct production affect byproduct supply. Joint costs (costs associated with production of both products) are borne by the main product and do not influence byproduct supply.

Byproducts are typically available at lower costs then the same product produced elsewhere as a main product, (e.g., REEs produced as a byproduct of iron ore in China would have lower production costs than would REEs produced elsewhere in the world as a main product).

Byproducts, typically, are not free goods, meaning that there are costs associated with their production. Byproducts could be without cost if two conditions are met: (1) production of main product must require the separation of the byproduct, and (2) no further processing of the byproduct is required after separation.

---

share joint production costs as no single co-product can support mine development costs alone.

Global Mineral Production

Table 2 provides data on the global production of critical minerals and the leading producing countries. The data shows that production for nearly all of the critical minerals has increased since 2000, many of which have doubled (e.g., chromium, indium, lithium, manganese, niobium, and tantalum) or tripled (e.g., cobalt, gallium, and tellurium) in the amount produced.

Table 2. Critical minerals: global production and leading producers, selected years (data in metric tons (mt) or million metric tons (m mt) unless otherwise noted)

Global Production					
Mineral	**2000**	**2010**	**2017**	**Leading Producers in 2017**	**Comments**
Aluminum (bauxite)	135.0 m mt	209.0 m mt	309 m mt	Australia (28.5%), China (22.6), Brazil (12.5%), Guinea (15%), Others (21.4%)	No bauxite produced in the United States
Antimony	118,000 mt	167,000 mt	137,000 mt	China (72%), Others (28%)	Relatively little U.S. production; none reported in 2017
Arsenic	33,900 mt	52,800 mt	34,600 mt	China (69%), Others (31%)	No U.S. production
Barite	6.2 m mt	7.85 m mt	8.7 m mt	China (37%), India (18%), Others (45%)	No U.S. production
Beryllium	280 mt	205 mt	210 mt	U.S. (71%), Others (29%)	U.S. is a net exporter
Bismuth	5,880 mt	8,900 mt	17,100 mt	China (73%), Others (27%)	No U.S. production
Cesium	NA	NA	NA	NA	No U.S. production
Chromium	14.4 m mt	23.7 m mt	30.2 m mt	South Africa (46.2%), Kazakhstan (12.9%), Others (40.9%)	No U.S. production
Cobalt	33,300 mt	89,500 mt	120 m mt	DRC (61%), Others (39%)	Some U.S. production as byproduct of copper

Table 2. (Continued)

Global Production					
Mineral	**2000**	**2010**	**2017**	**Leading Producers in 2017**	**Comments**
Fluorspar	4.5 m mt	6.0 m mt	5.7 m mt	China (61%), Mexico (18%), Others (21%)	No U.S. production
Gallium	100,000 kg	182,000 kg	320,000 kg	China (94%), Others (6%)	Small amount of low- grade gallium as a U.S. byproduct
Germanium	71,000 kg	118,000 kg	106,000 kg	China (57%), Others (43%)	Small amount of U.S. production as byproduct of zinc ore
Graphite (Natural)	571,000 mt	925,000 mt	897,000 mt	China (75%), Brazil (10%), Others (15%)	No U.S. production
Hafnium	NA	NA	NA	NA	See zirconium
Helium	98 million cubic meters (mcm)	75 mcm	160 mcm	U.S. (57%), Qatar (28%), Algeria (8.7%), Others (6.3%)	U.S. is a leading producer
Indium	335 mt	609 mt	714 mt	China (40%), South Korea (31.5%), Others (27.5%)	Data is for refinery production
Lithium	14,000 mt	28,100 mt	38,000 mt	Australia (58%), Chile (21%), China (9.8%), Argentina (8.3%), Others (2.9%)	Some U.S. production
Magnesium Metal	368,000 mt	757,000 mt	1.1 m mt	China (89%), Others (11%)	Some U.S. production
Manganese	7.28 m mt	13.9 m mt	17.3 m mt	South Africa (31%), Australia (16%), Gabon (12.7%), China (9.8%), Others (30.5%)	No U.S. production
Niobium	32,600 mt	62,900 mt	69,100 mt	Brazil (88%), Others (12%)	No U.S. production
Platinum	155,000 kg	192,000 kg	199,000 kg	South Africa (72%), Russia (11%)	The data in this row represents platinum only.

Global Production					
Mineral	**2000**	**2010**	**2017**	**Leading Producers in 2017**	**Comments**
					Palladium production of 225,000 kg is split between two major producers – South Africa (39%) and Russia (38%). Small amount of U.S. production
Potash	25.3 m mt	33.7 m mt	41.4 m mt	Canada (29%), Russia (17.6%), China (13%), Others (40.4%)	Relatively little U.S. production (roughly 1%)
Rare Earth Elements	83,500 mt	133,000 mt	132,000 mt	China (80%), Australia (14%), Others (6%)	No production in 2017. The USGS estimates U.S. production to be around 15,000 mt in 2018.
Rhenium	28,400 kg	47,200 kg	51,600 kg	China (55%), Poland (19%), U.S. (17%), Others (9%)	Relatively small amount of U.S. production as byproduct of copper recovery
Rubidium	NA	NA	NA	NA	No U.S. production
Scandium	NA	NA	NA	NA	No U.S. production
Strontium	520,000 mt	405,000 mt	255,000 mt	Spain (35.3%), Mexico (28%), China (19.6%), Iran (15.7%)	No U.S. production
Tantalum	836 mt	681 mt	1,810 mt	DRC (42%), Rwanda (24%), Nigeria (8.5%), Others (25.5%)	No U.S. production
Tellurium	125 mt	NA	470 mt	China (68%), Japan and Russia about 12% each, Others (9.2%)	Some U.S. production as byproduct of copper and lead recovery

Table 2. (Continued)

Global Production					
Mineral	**2000**	**2010**	**2017**	**Leading Producers in 2017**	**Comments**
Tin	238,000 mt	256,000 mt	313,000 mt	China (29.7%), Indonesia (26.5%), Burma (15%), Others (28.8%)	No U.S. production
Titanium	4.3 m mt	6.4 m mt	5.5 m mt	South Africa (18%), China (15%), Canada (16%), Australia (13%)	Relatively small amount of U.S. production
Tungsten	37,400 mt	68,800 mt	82,100 mt	China (82%), Others (18%)	No U.S. production
Uranium	NA	1,506 mt	1,021 mt	Kazakhstan (39%, Canada (22.5%), Australia (10%)	Some U.S. production
Vanadium	43,000 mt	57,600 mt	71,200 mt	China (56%), Russia (25%), South Africa (11.2%)	No U.S. production
Zirconium	1.04 m mt	1.25 m mt	1.55 m mt	Australia (32.5%), South Africa (24.3%), China (9%), Others (34.2%)	Some U.S. production

Source: USGS, Mineral Commodity Summaries, 2019. Data on uranium from the Energy Information Administration.

Notes:

kg = kilograms;

NA = not available.

DRC = Democratic Republic of Congo;

U.S. = United States.

The table is using 2017 data from the USGS Mineral Commodity Summaries, 2019 report because the report provides actual data for 2017 and only estimated data for 2018.

Some countries may be listed as leading producers but not listed as leading reserve holders of the same mineral listed in Table 4.

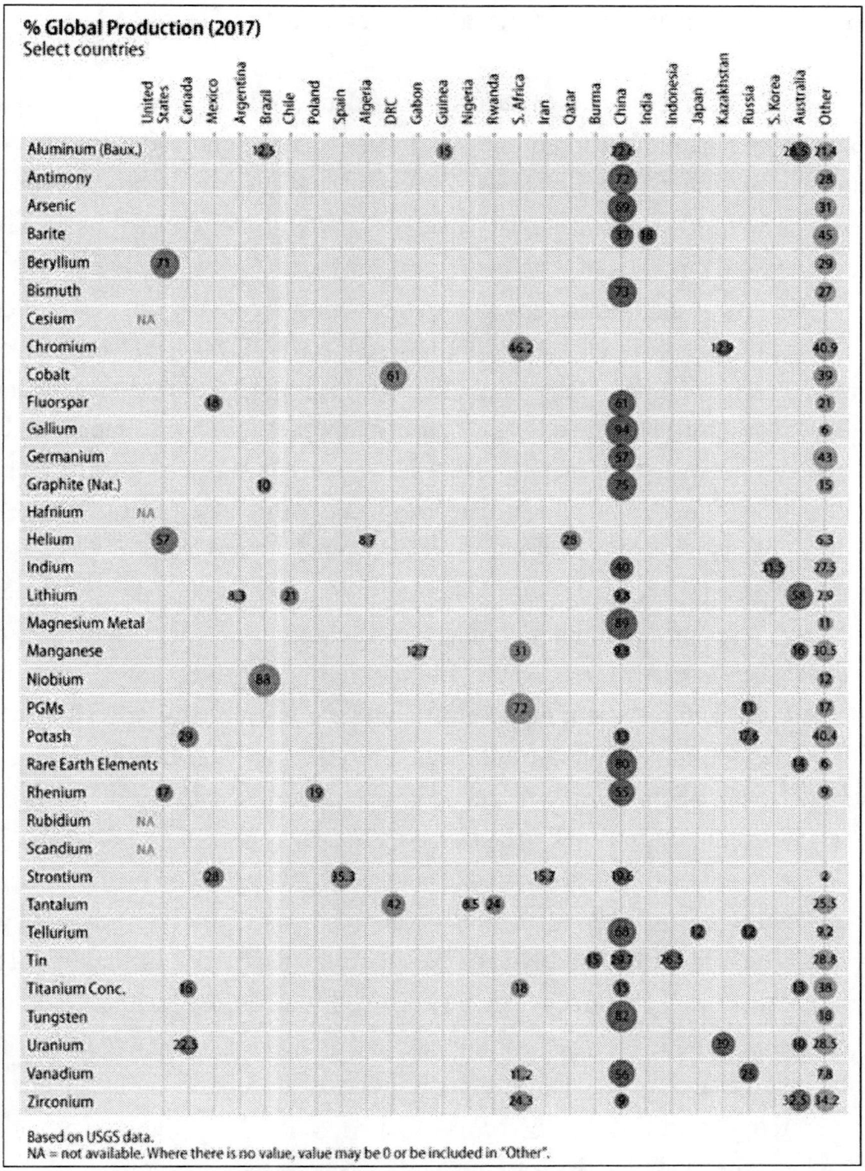

Source: Figure created by CRS based on USGS Mineral Commodity Summaries, 2019.

Notes: Color codes: Blue = North America; Purple = South America; Orange = Europe; Green = Africa &Middle East; Red = Asia and Russia; Dark Green = Australia; and Gray = Other countries that are not specifically mentioned in the previous columns.

Figure 1. Critical Minerals: Global Production (2017).

Secondary Recovery of Critical Minerals in the United States

Secondary recovery can occur from waste products during the metal refining and manufacturing process or from discarded end use products. As indicated in Table 3, in the United States, there is little to no production or reserves and little to no secondary recovery currently for many (but not all) of the critical minerals of high net import reliance.

There is a significant amount of secondary recovery in the United States of nine critical minerals according to the USGS Mineral Commodity Summaries: aluminum, chromium, cobalt, gallium, indium, magnesium metal, platinum group metals, tin, and titanium. While U.S. capacity for secondary recovery of metals and other materials has not grown much between 1997 and 2016, rates of recovery have fluctuated annually. Steel is the most recycled material in the United States. There are well established infrastructures, for old and new scrap, for selected metals such as steel, copper, aluminum, cobalt, and chromium.[53] For many other metals, such as manganese, REEs, and niobium, little-to-no recycling takes place in the United States because it is either economically or technically not viable.[54] Countries in the European Union, Japan, and South Korea are strengthening their efforts in secondary recovery as emerging markets (e.g., China and India) seek to secure greater access to primary materials.

The quantity of most metal and materials available for recycling will likely continue to meet a fraction of demand, particularly if demand is rising. The rate of availability (i.e., based on the useful life of the product) puts a limit on how much can be recycled. According to the National Research Council, the primary impediment facing secondary recovery in the United States is the lack of clear policies and programs at all levels of government to embrace the recovery of materials.[55] Without a national mandate, the National Research Council report indicates that state and

---

[53] Old scrap is material in discarded or obsolete products that have reached the end of their life. New scrap is material generated from processing primary materials. Almost all new scrap is recycled, and thus, it is not always considered a substitute for primary material.

[54] Manganese used in making steel is typically recycled as part of ferrous and nonferrous steel scrap.

[55] National Research Council, *Minerals, Critical Minerals, and the U.S. Economy*, Washington, DC, 2008.

local governments are likely to continue a "patchwork" of programs and policies.[56]

Table 3. U.S. secondary recovery of critical minerals, 2017

Mineral	Secondary Recovery as % of U.S. Apparent Consumption (unless otherwise noted)	Comments
Aluminum	28%	
Antimony	Unknown	Majority of U.S. supply is from secondary sources
Arsenic	None reported	
Barite	None reported	
Beryllium	20%-25%	
Bismuth	<5%	Both old and new scrap
Cesium	Unknown	Some formate brines reprocessed
Chromium	29%	
Cobalt	29%	No primary production; secondary recovery of purchased scrap
Fluorspar	Unknown	Very little
Gallium	Unknown	No old scrap, significant new scrap recovered
Germanium	NA	About 30% worldwide
Graphite	Unknown	Not much because of raw material abundance
Hafnium	Negligible	
Helium	NA	Very little
Indium	Significant domestic recycling but amount not known	On a global scale, secondary production greater than primary production
Lithium	Very little	DOE grant was awarded in 2009 for a recycling facility. A U.S. recycling facility for lithium-ion vehicle batteries opened in 2015.
Magnesium metal	120,000 tons	Old and new scrap
Manganese	Negligible	
Niobium	none reported	May be as high as 20% according to USGS.
Platinum	Known for platinum only	120,000 kilograms of platinum group metals recovered globally from old and new scrap
Potash	None	
REEs	Very little	
Rhenium	Some	

[56] Ibid.

Table 3. (Continued)

Mineral	Secondary Recovery as % of U.S. Apparent Consumption (unless otherwise noted)	Comments
Rubidium	None	
Scandium	None	
Strontium	None	
Tantalum	New scrap recovered but amount unknown	May be as much as 10% according to USGS.
Tellurium	Very little	
Tin	25%	12,300 tons, mostly old scrap
Titanium	69,600 tons scrap metal	
Tungsten	NA	Old and new scrap
Uranium	NA	
Vanadium	NA	Significant amount from spent chemical process catalysts
Zirconium	Some	

Source: USGS Mineral Commodity Summaries, 2019.

Notes: NA = not available. Unknown = no data reported by the USGS. The table is using 2017 data from the USGS Mineral Commodity Summaries, 2019 report because the USGS 2019 summaries provides actual data for 2017 and only estimated data for 2018.

Table 3 illustrates the point that there is very little secondary recovery of critical minerals and metals in the United States.[57] The data could indicate that there is a lack of infrastructure for secondary recovery of critical minerals and metals. Economic and technological factors must also be evaluated as to whether the benefits outweigh the costs for recovering certain materials, particularly the small amounts of critical minerals that may be available for secondary recovery (from manufacturing waste or end use products). Additional R&D may be needed to determine whether secondary recovery of the most import-dependent minerals could be increased to reduce U.S. import reliance.

In 2018, the USGS reports that for base metals and precious metals the recycling rate is much different. For example, the recycling rates were 28% for aluminum, 35% for copper, 52% for nickel, 18% for silver, and 25% for zinc. In 2014, steel in the auto industry was recycled at 106%—more

[57] Base metals are any of the nonprecious metals. Precious metals include gold, silver, and platinum.

steel than was used for domestic manufacturing. The recycling rate of steel is 90% for appliances containing steel and 67% for steel cans.

Reserves and Resources

There is a distinction between what is described when using the terms *reserves* and *resources* in the context of minerals. Reserves are quantities of mineral resources anticipated to be recovered from known deposits from a given date forward. All reserve estimates involve some degree of uncertainty. Proved reserves are the quantities of minerals estimated with reasonable certainty to be commercially recoverable from known deposits under current economic conditions, operating methods, and government regulations. Current economic conditions include prices and costs prevailing at the time of the estimate. Estimates of proved reserves do not include reserves appreciation.

Resources are concentrations in the earth's crust of naturally occurring minerals that can conceivably be discovered and recovered. Undiscovered technically recoverable resources are minerals that may be produced as a consequence of natural means, or other secondary recovery methods, but without any consideration of economic viability. They are primarily located outside of known deposits.

U.S. Critical Mineral Reserves and Resources

Regarding reserves, the USGS lists little to no reserves in all 35 of the critical minerals except for helium and beryllium and significant resource potential in only tungsten, lithium, vanadium, uranium,[58] and REEs. Of the 14 critical minerals listed as 100% import dependent, the USGS lists some reserves for two: REEs and vanadium (see Table 4 and Figure 2).[59]

---

[58] Uranium data obtained from the Energy Information Administration.

[59] For example, note the following minerals with some reserves and or production: arsenic—small reserves; asbestos— small reserves; bauxite—small production and reserves; fluorspar—byproduct of lime; thorium ore—some reserves, no U.S. production; yttrium—some reserves, little production; rare earths—some production, large reserves; rubidium and thallium—no production, small reserves. China produces 99% of the world's yttrium. There

Regarding resources, USGS identifies some resource potential for cesium, manganese, and niobium. There are byproduct resources of cobalt, germanium, tellurium, and rhenium that are associated with main products such as copper, zinc, and bauxite (see Table 4). The USGS is uncertain about U.S. and global reserves of several critical minerals as not enough data are available according to the USGS.[60]

Global Critical Mineral Reserves and Resources

According to the USGS, at the global level, there are significant or abundant resource potential for the critical minerals for which the agency has data, which is some but not all of the critical minerals.

Table 4. Critical minerals: global resources and reserves, 2017 (data in metric tons unless otherwise noted)

Mineral	Resources	Reserves	Leading Reserve Holders by Country	Comments
Aluminum (Bauxite)	Abundant global resources; U.S. resources not significant	30 b mt	Guinea (24.6%), Australia (20.6%), Vietnam (12.3%), Brazil (8.6%), Jamaica (6.6%)	China has 3% of reserves but produces almost 23% of bauxite.
Antimony	Some resource potential in Alaska, Montana and Idaho. Principal global resources in Australia, Bolivia, China and Mexico	1.5 b mt	China (32%), Russia (23%), Bolivia (21%)	The United States has about 4% of global reserves
Arsenic	Unknown	NA	NA	No U.S. reserves; world reserves unavailable but estimated at about 20x current global production.

is no planned or current production of bauxite or fluorspar on public lands. The USGS indicates U.S. mineral reserves of beryllium, helium, lithium, platinum group metals, REEs, tungsten, and uranium.

[60] See the USGS 1802 Study and Mineral Commodity Studies.

Mineral	Resources	Reserves	Leading Reserve Holders by Country	Comments
Barite	2 billion tons worldwide; significant U.S. resources	320 m mt	Kazakhstan (26.5%), India (16%), China (11%), Turkey (11%), Others (35.5%)	No U.S. reserves
Beryllium	60% of world's estimated100,000 mt of resources in the U.S.	NA	NA	
Bismuth	NA	NA	NA	No U.S. reserves
Cesium	Some U.S. resource potential, world resources unknown	90,000 mt	Zimbabwe (67%), Namibia (33%)	No U.S. reserves
Chromium	Small U.S. resources Significant world resources	560 m mt	Kazakhstan (41%), South Africa (35.7%), India (17.8%), Others (5.5%)	
Cobalt	Small U.S. resources 25 m mt terrestrial; 120 m mt seabed nodules	6.9 b mt	DRC (49%), Australia (17.4%), Cuba (7.2%), Others (26.4%)	
Fluorspar	500 million tons worldwide; significant resources in phosphate rock in the United States	310 m mt	Mexico (21.9%), China (13.5%), South Africa (13.2%), Others (51.4%)	No stand-alone U.S. reserves, but significant amounts contained in phosphate rock
Gallium	Significant resources worldwide in bauxite and zinc but only 10% recovered; sub-economic resources in the U.S. contained in bauxite	NA	Unknown	
Germanium	Uncertain	NA	Unknown	
Graphite	>800 m mt inferred resources. Small U.S. resources	300 m mt	Turkey (30%), China (24.3%), Brazil (24%), Others (21.7%)	
Hafnium	NA	NA	NA	
Helium	20,600 million cubic meters in the United States	NA	U.S., Algeria, Russia	U.S. is a world leader in reserves with 3,900 million cubic meters
Indium	NA	NA	NA	NA
Lithium	47 m mt globally; 6.9 m mt in the United States	14 m mt	Chile (57%), Australia (19.3%), Argentina (14.3%) China (7%), Others (2.4%)	Small U.S. reserves but significant resources

Table 4. (Continued)

Mineral	Resources	Reserves	Leading Reserve Holders by Country	Comments
Magnesium compounds	Billions of tons worldwide	8.5 b mt	Russia (27%), North Korea (27%), China (11.8%), Others (34.2%)	
Manganese	Low grade resources in the United States; 78% of global resources in South Africa	760 m mt	South Africa (30.2%), Ukraine (18.4%), Brazil (14.5%), Australia (13%), Others (23.9%)	
Niobium	Resources more than adequate supply to meet global demand; Low grade resources in the U.S.	9.1 m mt	Brazil (80%), Canada (17.6%), Others (2.4%)	
Platinum Group Metals	100 million kilograms	69,000 mt	South Africa (91%)	Some U.S. reserves. Most of the world's resources are in South Africa
Potash	7 billion tons in the United States, 250 b mt worldwide	NA	Canada, Belarus, Russia	USGS did not report total world reserves
Rare Earth Elements	Abundant but not always in minable concentrations; significant resources in the United States	120 m mt	China (37%), Brazil (18%), Russia (15%), India (5.8%), Australia (2.8%)	Some U.S. reserves
Rhenium	Significant U.S. and world resources	2,400 mt	Chile (54%), United States (16.6%), Russia (12.9%), Others (16.5%)	
Rubidium	Significant world resources	90,000 mt	Namibia (55%), Zimbabwe (33%), Others (12%)	
Scandium	Abundant world resources		Unknown	
Strontium	About 1 billion tons	6.8 b mt	China (percent of total unknown)	USGS did not report reserve data for other countries
Tantalum	Some resources in the U.S.	>110,000 mt	Australia (70%), Brazil (30%)	Data unavailable for other countries, even though 80% of production is in Africa

Mineral	Resources	Reserves	Leading Reserve Holders by Country	Comments
Tellurium	NA	31,000 mt	China (21.3%), United States (11.3%), Others (67.4%)	Some U.S. reserves contained in copper and lead ores
Tin	Abundant worldwide; some resources in the U.S., mostly in Alaska	4.7 m mt	China (23.4%), Indonesia 17%), Brazil (15%), Others (44.6%)	No U.S. reserves reported
Titanium	2 billion mt of titanium mineral concentrate worldwide	880 m mt	Australia (28.4%), China (26.1%), India 9.7%), Others (35.8%)	Data in this row does not include rutile (a related mineral). Small amount of U.S. reserves
Tungsten	Abundant global resources; the United States has significant tungsten resources	3.2 m mt	China (57.5%), Others (42.5%)	
Uranium	7,641,600 tons worldwide	4.4 m mt	Australia (26%), Canada (11%) Kazakhstan (8.2%), Niger (7.2%), Namibia (6.8%), Russia (6.2%)	
Vanadium	63 million tons worldwide; significant resources in the United States	20 m mt	China (47.5%), Russia (25%), South Africa (17.5%), Others (10%)	Small U.S. reserves
Zirconium	Substantial zirconium resources as part of titanium and phosphate rock	73 m mt	Australia (57.5%), South Africa (19%), Others (23.5%)	

Source: USGS, Mineral Commodity Summaries, 2019. Data on uranium from the Energy Information Administration, 2018 Domestic Uranium Production Report, May 2019.

Notes: mt = metric tons; m mt = million metric tons; kg = kilograms; b mt = billion metric tons; NA = not available.

Global resource potential is either unknown or uncertain for bismuth, cesium, germanium, indium, and tellurium. Most of the germanium, indium, and tellurium are obtained as byproducts of base metal production. China leads the world in reserves in seven critical minerals, including antimony, REEs, strontium, tellurium, tin, tungsten, and vanadium (see Table 4). China is among the top three reserve holders in barite, fluorspar, graphite, magnesium compounds, and titanium.

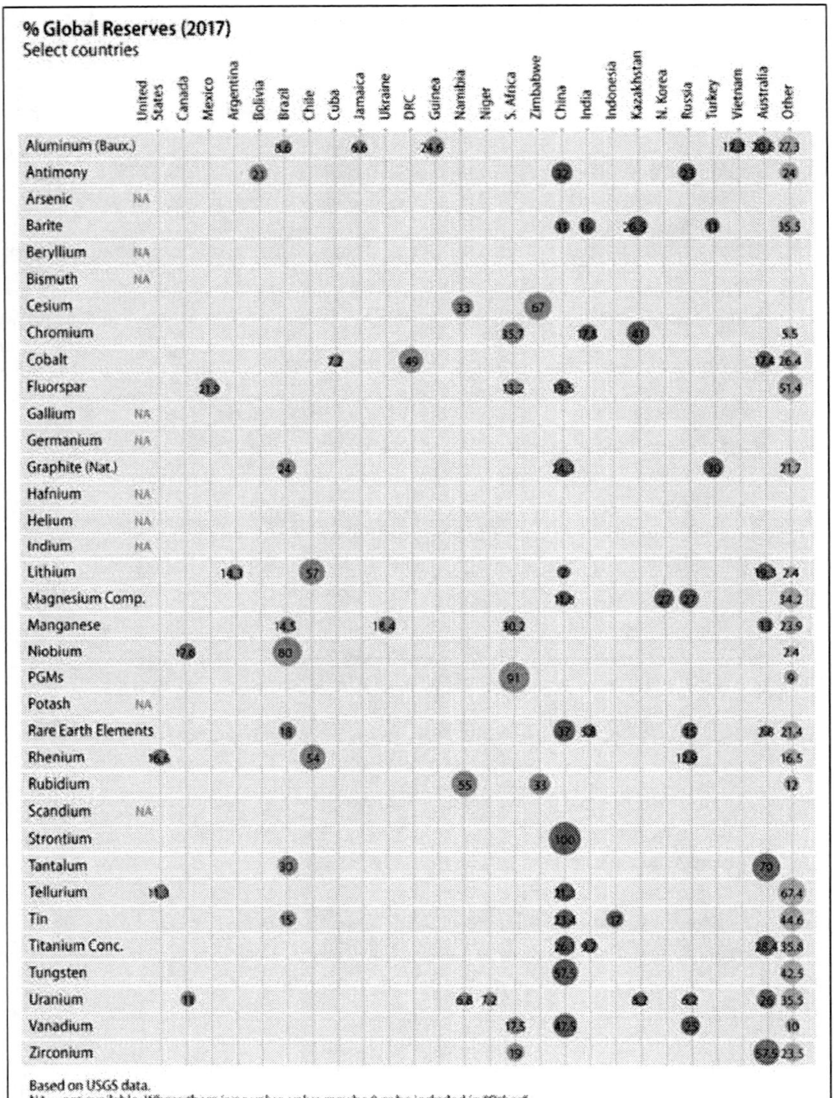

Source: Figure created by CRS based on USGS data, Mineral Commodity Summaries, 2019.

Note: Color codes: Blue = North America; Purple = South America; Orange = Europe; Green = Africa; Red = Asia and Russia; Dark Green = Australia; and Gray = Other countries that are not specifically mentioned in the previous columns. USGS reports Strontium reserve data only for China.

Figure 2. Critical Minerals: Global Reserves (2017).

Table 4 provides available information on global resources of critical minerals, as well as information on the size of the reserves. Figure 2 provides information on the regional distribution of the reserves.

Mineral Exploration

Exploration expenditures for minerals in the United States have been rising since 2001. The United States has maintained about 8% of the annual exploration budget for minerals worldwide from 1997 to 2017. In 2017, these expenditures in the United States were at 225 exploration sites (out of 2,317 exploration sites worldwide); 41% of the U.S. sites were in Nevada, 14% in Alaska, and 11% in Arizona.[61] It can take many years for mining firms to find and bring an economic deposit into production. Thus, it is important for the industry to keep mineral projects in the exploration-development process.

In general, mineral exploration in the United States remains focused on a few minerals, most of which not considered critical. Exploration activity in the western states is primarily for gold, copper, molybdenum, silver, tungsten, and uranium. There had been some reported interest in expanding silica sand operations in Nevada, developing a copper-cobalt-gold project in Idaho on Forest Service land,[62] and thorium production on federal lands along the Idaho/Montana border.

Globally, Canada leads with the most active exploration sites, mostly for gold and base metals (over 500 sites), followed by Australia (about 500 sites) with investments mostly in gold, base metals, and uranium.

Locations and Minerals Being Explored

The locations and minerals being explored can be shape how critical mineral supply chains are or may evolve. These supply chains have

---

[61] SMME, *Annual Mining Review*, p. 47.

[62] 71 *Federal Register* 64237, "Idaho Cobalt Project Plan of Operations, Salmon Challis National Forest, Lemhi County, ID," November 1, 2006, http://www.govinfo.gov/app/details/FR-2006-11-01/E6-18362.

relevance to various policy questions, including what is the long-term investment strategy in the United States to develop mineral extraction and downstream metal and manufacturing capacity; and, if the focus is on building a reliable supply chain, what part of that supply chain makes sense to develop in the United States?

There have been recent new additions to the annual USGS mineral exploration review. Data on lithium, niobium, rare earth elements, and tungsten are now included. Data for other minerals such as scandium, vanadium, and yttrium have been compiled since 2014.

The big global exploration story is about lithium. In 2016, global exploration dollars for lithium, cobalt, and gold rose significantly. The lithium exploration expenditures increased four-fold since 2015 and active exploration sites rose from 56 in 2012 to 167 sites in 2017. Lithium exploration expenditures, for example, rose from $22 million in 2015 to $128 million in 2017 as the number of lithium exploration companies grew from 23 in 2015 to 125 in 2017. The price of lithium rose by more than 150% from 2007 to 2016 and sits at 83% higher than its 10-year average. The number of cobalt sites rose by 121% since 2016.[63]

In the United States in 2017, gold remains in the top spot for the number of exploration sites (47%) followed by copper (12%), then lithium with 7% of the sites. USGS noted that there is continued interest in graphite, REEs, and tungsten in the United States, but the most notable sites are in gold exploration. Overall, 54% of the sites actively explored in the United States are for gold and silver and 22% for base metals. Worldwide, gold or silver accounts for 84% of the sites actively explored.[64]

The USGS reported that the United States has accounted for about 7% to 8% of overall global exploration budget over the past 10 years (about $611 million in 2017). However, the annual review is not exactly a country-by-country comparison because the USGS uses regions such as Latin America and Africa to compare with individual countries such as Canada, Australia, and the United States. The mineral exploration budget

---

[63] SMME, *Annual Mining Review*, p. 35.
[64] Ibid, p. 40.

directed at U.S. mineral deposits is above that of China (5%), Russia (4%), and many countries in Latin America.[65]

Latin America attracts the most exploration dollars with $2.4 billion, most of which are for gold and silver (58%) followed by base metals at 22% of exploration expenditures. Chile has seen the most investment in Latin America, followed by Peru. Latin America is home to 70% of the world's known lithium deposits, known as the "lithium triangle" consisting of Chile, Argentina, and Bolivia. In Argentina, lithium exploration sites account for 44% of exploration expenditures followed by gold/silver at 42%, and copper at 9%. Lithium is most developed in Chile because of its superior infrastructure for mining. Most exploration projects in Chile are for copper (49%) and gold (29%).[66]

There has been an uptick in lithium exploration in Australia as well. China invested $650 million (in U.S. dollars) in Australia in 2016, looking for lithium and gold, primarily.[67] As ore grades decline at known reserve locations, many exploration companies are searching for high-grade deposits in remote locations, including the ocean floor.

DEMAND: CRITICAL MINERAL END USES AND U.S. IMPORT RELIANCE

Demand for Critical Minerals

The demand for mineral commodities is a derived demand which differs from consumer goods demand. Minerals are used as inputs for the production of goods and services. For example, the demand for rare earth elements is derived from the production of their end-use products or use, such as flat panel displays, automobiles, or catalysts. As a result, the demand for critical minerals depends on the strength of the demand of the final products for which they are inputs. An increase in the demand for the

---

[65] Ibid, p. 49.

[66] Ibid, pp. 40-43.

[67] Ibid, p. 50.

final product will lead to an increase in demand for critical minerals (or their substitutes).

In the case of derived demand, when mineral and metal prices rise, the extent to which the quantity of a material declines depends largely on the degree to which its price increase can be passed on to the final consumer, as well as the proportion of the final good's price that is accounted for by the mineral/metal commodity. That is, it might depend on the amount of critical mineral or metal used per unit of output. The major variables that determine the growth in demand for consumer goods are price and income growth.[68]

U.S. and Global Demand

U.S. demand has declined for some critical minerals, and for others, demand has increased but not as much (in relative terms) as the increase in global supply. For example, over the past 20 years consumption fell for aluminum, chromium, manganese, platinum group metals, REEs, titanium, and tantalum, among others, and demand grew slowly for lithium, germanium, and graphite. Only for tellurium, niobium, and indium did the United States experience rapid demand growth (relative to supply).[69] Some of the demand drivers in recent decades for critical minerals include permanent magnets using REEs, batteries using cobalt and lithium, automobiles and electronics using tantalum and niobium, and vanadium for steel production.

Global demand data for each of the minerals listed as critical were not available at the time of this writing. Global demand data could shed more light on where the minerals are being used for metal alloying, the manufacturing of component parts, and final products. Embodied metals (those that are imported as final products) are not counted as demand.[70]

Many critical minerals, (e.g., manganese, tungsten, and vanadium) are used for steelmaking and infrastructure projects, such as roads, housing,

---

[68] Gary A. Campbell, "Theory of Mineral Demand," *Economics of the Mineral Industries*, American Institute of Mining, Metallurgical, and Petroleum Engineering, 1985.

[69] USGS Mineral Commodity Summaries, Various Years, 1997–2019.

[70] Embodied minerals/metals are those embedded in the end use product such as an imported laptop or automobile.

rail lines, and electric power grids. Others (e.g., REEs, lithium, indium, tantalum, gallium, and germanium) are used in the manufacturing of high-value electronic products, such as laptops and batteries, renewable energy systems, and other consumer goods, such as automobiles and appliances (see Table 5).

Demand for Critical Minerals in China

There has been a surge in demand for critical minerals in China. China's demand for natural resources rose to historic levels and may continue to rise over the long term, even with a slowing economy. In the recent past, China has been the fastest growing market for niobium, and in 2010 accounted for 25% of world niobium consumption.[71] Manganese consumption rose from about 2,200 metric tons (mt) in 2003 to about 9,000 mt in 2008.[72] China's demand for vanadium paralleled that of steel demand and rose 13% annually from 2003 to 2009. In general, vanadium demand in China is projected to double from 2010 to 2025 because of its continued use in steelmaking (including new steel-hardening requirements) and because of the potential for application in new battery technology used for large-scale renewable energy storage (e.g., vanadium-redux flow battery-VRFB).[73] In 2010, China accounted for 85% of chrome ore import demand[74] and is the world's leading producer of steel (accounting for over half the world's production in 2017 based on the most recent data).[75] Chromium is a major production input for stainless steel. China's chrome imports will likely continue to increase as stainless steel demand at the global level remains a big part of China's high-valued exports, urbanization, and future industrial practices.

---

[71] IAMGOLD Corporation, "Niobium 101," March 28, 2012.

[72] Shaw River Manganese Limited, "Manganese Fact Sheet," 2010, http://www.shawriver.com.au.

[73] Schauss, Steven, "A Bull Market Storm Brewing for Vanadium," January 2, 2019, http://www.mining.com/web/bull- market-storm-brewing-vanadium.

[74] International Chromium Development Association, "Industrial Minerals," *Mining Engineering*, June 2011.

[75] USGS, Mineral Commodity Summaries, 2019.

Overall, in 2017, China's cobalt smelters accounted for 60% of global supply, and 77% of cobalt demand in China went into batteries.[76] In 2017, China accounted for about 25% of platinum demand, primarily used in jewelry making, and 26% of palladium demand, much of which is used in catalytic converters in automobiles.[77]

In order for this increasing demand scenario in China to play out, the cities would need to fill up with enough people who are making high enough wages to support the economic growth that China is seeking. It is uncertain whether such a high level of consumer demand will materialize. China's economic growth has slowed considerably in the recent past from around 10% annually in the first decade of the 2000s, to around 6% in 2014.[78] However, China's demand for minerals will continue to put pressure on U.S. access to reliable supplies.

U.S. Imports of Strategic and Critical Minerals

Aside from a small amount of recycling, the United States is 100% import reliant on 14 minerals on the critical minerals list, minerals that provide critical support for the U.S. economy and national security such as, graphite, manganese, niobium, rare earths, and tantalum, among others. The United States is more than 75% import reliant on an additional 10 critical minerals, including antimony, barite, bauxite, bismuth, potash, rhenium, tellurium, tin, titanium concentrate, and uranium.

The United States has increased its mineral imports from China over the past 20 years. Although the United States has diversified its sources for some of its material requirements since 1997, the United States imports significant quantities of critical minerals and metals and is dependent on

---

[76] "Global and China Cobalt Market Report, 2018-2023 Featuring 5 Chinese and 14 Global Manufacturers," News Provided by Research and Markets, November 9, 2018, https://www.prnewswire.com/news-releases/global-and-china- cobalt.

[77] Johnson Matthey, *PGM Market Report*, February 2018.

[78] Wright, Andrew, "This Is How China's Economy Has Changed in the Last 10 Years," World Economic Forum, June 22, 2016.

China as either a primary or major provider of raw materials and several metals as of 2017 (see Table 5 and Figure 3).

Table 5. Critical minerals: major end uses and net U.S. import reliance

Mineral	Major End Uses	Import Reliance (%)	Major Sources	Comments
Aluminum (Bauxite)	transportation, packaging, building, electrical	>75	Jamaica (46%), Brazil (25%), Guinea (15%), Other (14%)	The data reflect the import reliance for bauxite, the source mineral for aluminum
Antimony	ceramics, glass, and rubber products, fire retardant	85	China (61%), Other (39%)	Major sources are for antimony oxide
Arsenic	lead storage batteries, herbicides, insecticides, military applications	100	China (91%)	Import of arsenic metal
Barite	filler, extender, and weighing agent in paint, plastics and rubber	86	China (63%), India (14%), Others (23%)	
Beryllium	auto and consumer electronics, defense applications	17	Kazakhstan (44%), Japan (14%), Others (42%)	
Bismuth	additives for lead-free pipe fittings	97	China (80%), Others (20%)	
Cesium	photoelectric cells, and energy conversion devices	100	Canada	According to USGS, Most imports are from Canada, but percentage from Canada unavailable
Chromium	transportation, packaging, building, electrical	71	South Africa (97%)	Import reliance for chromite ore
Cobalt	super alloys, aircraft engines, batteries, permanent magnets	69	Norway (18%), China (12%), Japan (12%), Others (58%)	These imports reflect cobalt contained in metal, oxides and salts

Table 5. (Continued)

Mineral	Major End Uses	Import Reliance (%)	Major Sources	Comments
Fluorspar	used in processing aluminum, and uranium	100	Mexico (69%), Vietnam (10%), South Africa (8%), Other (13%)	
Gallium	integrated circuits (in high-tech equipment), light emitting diodes (LEDs), solar cells	100	China (32%), UK (28%), Germany (15%), Ukraine (14%), Other (11%)	
Germanium	fiber optics, infrared optics, solar cells, other solar energy applications	>50	China (58%), Belgium (26%), Other (14%)	Import reliance for germanium metal
Graphite (Natural)	steelmaking, refractory applications, foundry operations, brake linings	100	China (37%), Mexico (29%), Canada (17%), Other (17%)	
Hafnium	super alloys	NA	Germany, France, UK	Percentage from each country unavailable
Helium	lifting gas, lab applications, MRI, welding	—		United States is a net exporter
Mineral	Major End Uses	Import Reliance (%)	Major Sources	Comments
Indium	electrical conduction, liquid crystal displays (LCDs), solar cells and photovoltaics	100	China (27%), Canada (22%), Other (51%)	
Lithium	rechargeable batteries, ceramics, glass, chemical compounds	>50	Argentina (51%), Chile (44%), Others (4%)	
Magnesium Compounds	Agriculture, chemicals, construction, and industrial applications	51	China (57%), Canada (22%), Others (21%)	
Manganese	production of steel and other metals	100	Gabon (74%), South Africa (13%), Australia (8%), Others (5%)	

Mineral	**Major End Uses**	**Import Reliance (%)**	**Major Sources**	**Comments**
Niobium	steel and super alloys	100	Brazil (72%), Canada (18%), Others (10%)	Imports of niobium include ore and concentrate, niobium oxides, ferroniobium, and niobium metal
Platinum Group Metals	auto catalysts, fuel cells, jewelry	71	South Africa (44%), Germany (15%), UK (10%). Others (31%)	This row represents platinum only. The United States is 38% import reliant on palladium most of which comes from Russia and South Africa
Potash	fertilizer, chemical industry applications	92	Canada (84%)	
Rare Earth Elements	permanent magnets, petroleum refining, glass, lasers, steel alloys, fluorescent lighting	100	China (80%)	
Rhenium	super alloys in high temperature turbine engine components and petroleum-reforming catalysts	81	Kazakhstan (34%), Canada (19%), South Korea (13%), Germany (10%), Others (24%)	
Rubidium	biomedical research, electronics, specialty glass	100	Canada	Percentage from Canada unavailable
Scandium	Ceramics, electronics, lasers, radioactive isotopes, lighting	100	Mostly from China, Europe, Japan, and Russia	Percentage from each country unavailable
Strontium	additive in drilling fluids for oil and gas wells	100	Mexico (52%), Germany (39%), Others (9%)	
Tantalum	capacitors for electronic devices	100	Brazil (35%), Rwanda (31%), Australia (15%), Others (19%)	

Mineral	Major End Uses	Import Reliance (%)	Major Sources	Comments
Tellurium	photovoltaic panels, solar cells, thermoelectric devices	>75	Canada (66%), China (27%), Others (7%)	
Tin	Chemicals, tinplate, solder and alloys	76	Indonesia (23%), Malaysia (23%), Peru (22%), Bolivia (17%), Others (15%)	
Titanium Concentrate	aerospace applications	92	South Africa (35%), Australia (27%), Canada (12%), Mozambique (11%), Others (15%)	
Tungsten	cutting tools, wear-resistant materials used in construction and metal making	>50	China (32%), Germany (9%), Bolivia (9%), Canada (8%), Others (42%)	
Uranium	fuel for nuclear reactors	93%	Canada, Australia, Russia	The United States supplied 7% of the uranium purchased by U.S. power plants in 2017.
Vanadium	steelmaking, aerospace applications	100	South Africa (46%), Russia (18%), Brazil (13%), China (10%), Others (13%)	
Zirconium	Used in ceramics, foundry sand, refractories, and abrasives	—	South Africa (59%), Australia (22%), Senegal (14%)	The United States is a net exporter

Source: USGS, Mineral Commodity Summaries, 2019.

Note: > = greater than.

While import reliance may be a cause for concern (and high levels of import reliance potentially a security risk), high import reliance is not necessarily the best measure, or even a good measure, of supply risk. A more relevant measure may be the reliability of the suppliers. The supply risk for potash or bauxite, for example, may not be the same as that for REEs or niobium due to the multiplicity of potential sources. There are a number of factors that affect the availability of mineral supplies that may have little to do with import reliance.

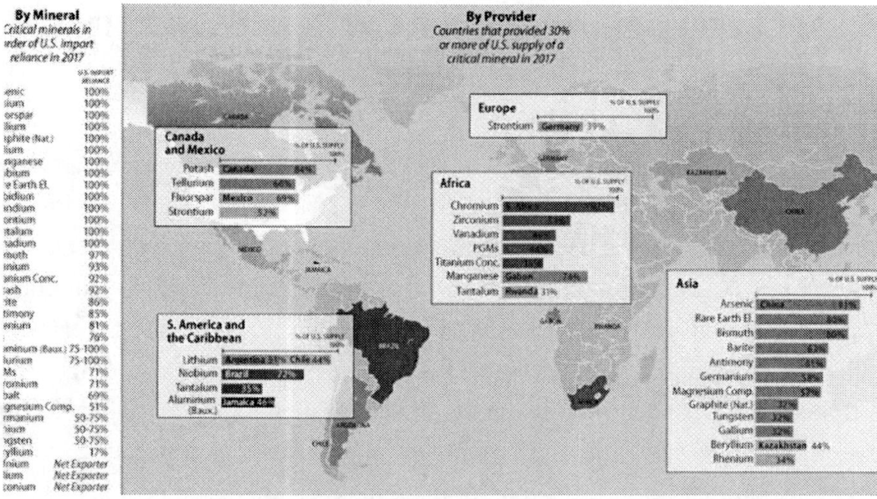

Source: Figure created by CRS based on USGS Minerals Commodities Summaries data, 2019.

Note: Countries listed in the bar graph represent the leading supplier of U.S. imports.

Figure 3. Critical Minerals: Net U.S. Import Reliance (2017).

A company that is the sole supplier, or a single country as a primary source, with export restrictions, would likely constitute supply risks. But any number of bottlenecks that might arise among both domestic and foreign producers, such as limited electric power, skilled labor shortages, equipment shortages, labor unrest, weather or transportation delays, and opposition on environmental policy grounds, could also pose supply risks. Any of these above-mentioned potential supply disruptions could raise costs or prices, and exacerbate the tightness of supplies. For other minerals, such as iron ore and molybdenum, the United States is self-sufficient. For aluminum, uranium, potash, cesium, and rubidium, the United States' chief trading partner is Canada, a stable ally. Also, U.S. companies have invested in overseas operations—for example, copper and bauxite mines—and, thus, U.S. supply sources for some materials are diversified, of higher quality, or lower cost, and located in countries that have extensive reserves and production capacity. Such conditions may not always exist in the United States, even when resources are present.

MATERIALS ANALYSIS OF CRITICAL MINERALS CONTENT IN FINISHED PRODUCTS AND SYSTEMS

Materials analysis is a useful tool to better understand various aspects of mineral demand. For example, such analysis can provide information on how material inputs are used in component parts and how components are used in larger systems such as solar arrays, wind turbines, and automobiles. Using a material analysis, an analyst can obtain information on the material intensity of a unit of production.[79] This analysis can lead to manufacturing efficiencies (i.e., getting the same or better performance using fewer materials) or show where and how material substitution, if possible, could occur. Manufacturing firms could then make short-term or long- term adjustments to their production processes.[80]

Even with materials efficiencies, where less metal is used per unit of output, overall demand growth and lack of short-term supply capacity often drives up mineral prices.[81] For example, households in some countries are likely to have multiple units of a variety of products such as laptops, flat panel televisions, and cell phones, etc. And because the materials intensity (small amounts per unit output) of critical minerals is relatively low for most end-use applications, low-cost manufactured goods may contain some high-cost materials.

The remainder of this section of the report provides information on the materials content of lithium-ion batteries, solar energy arrays, wind technologies, and permanent magnets, with a more detailed discussion of the material requirements for wind and solar energy systems.

---

[79] Material intensity is the measure of the mineral input per unit of output in energy or in units over time.

[80] Short-term adjustments are adjustments to production and do not require any major capital investment; long-term adjustments require major capital investment in equipment or facilities.

[81] David Humphreys, "The Great Metals Boom: A Perspective," *Resources Policy*, v. 35, 2010.

Lithium-Ion Batteries

The use of lithium-ion batteries for the rapidly growing electric vehicle market is expected to transform the material requirements for battery technology. Material analysis of lithium-ion batteries would bring to light useful insights on materials composition, cost, technologies, and supply chains. In the case of the lithium-ion (li-ion) battery[82] for electric vehicles, what is the material composition of the battery?[83] In other words, how much cobalt, lithium, nickel, and other materials are needed per battery, how much are the material costs for each battery, and what percent of the total battery manufacturing cost do the materials represent? Then, further, what is the battery cost per electric vehicles? Analysts would want to know the point at which material price increases would warrant a shift in the use of those materials. Other useful insights in materials analysis would be to understand the suite of battery technologies being developed, their manufacturing capacity, and the ownership structure of the supply chain for the materials and the batteries.

A 2017 study by a group of battery technology researchers examined the supply risks associated with lithium-ion batteries and other battery technologies to examine the implication for a carbon- reduced environment.[84] The authors posed the question: What are the material requirements for the battery? They identified features of a li-ion battery, e.g., low cost, high energy, and long life. They examined the raw material requirements for li-ion batteries, secondary supply potential, and supply risks associated with an exhaustible resource (e.g., mineral extraction may become uneconomic), the structure of the industry (e.g., whether there is a cartel or a monopoly producer involved), and a surge in demand.

---

[82] Helbig, et al., "Supply Risks Associated with Lithium-ion Battery Materials," *Journal of Cleaner Production*, October 12, 2017 (hereinafter referred to as Helbig 2017).

[83] Material composition of a product (MCP) is a unit of measurement used to study impact of metal/minerals on demand for traditional material. MCP measures the efficiency of converting raw materials into final end use products. The greater the efficiency, the less demand for the material per unit of output.

[84] Helbig 2017. The li-ion battery is used in consumer electronics such as cell phones, laptops, notebooks, power tools, electric vehicles, and grid storage.

They used supply risk indicators discussed earlier, such as the risk of supply reduction, the risk of a surge in demand, market concentration, political stability, substitutability, and recyclability.

The researchers' second step was to determine the supply risk score on the technology level, for each of the six battery types.[85] There is a lithium-cobalt oxide battery which has a high energy density but also a high cobalt content and price. The steep country risk associated with cobalt production in the Democratic Republic of the Congo (DRC) led researchers to look for alternative suppliers and materials that would provide high energy density and long life with less or no cobalt. One example would be to use a manganese-oxide battery, wherein cobalt is partially replaced by nickel and manganese.

They pointed out that there are several new battery types that use combinations of lithium, aluminum, cobalt, iron, nickel, copper, graphite, phosphate, titanium, and manganese. The researchers identified lithium as needed for all battery types and graphite used for all except the lithium-iron-phosphate (LFP-LTO) type, which uses titanium instead. They reported that with a market breakthrough (by 2035) in the use of electric vehicles containing lithium battery technology, an annual growth rate of 7.5% is needed for lithium supply and 3% growth rate in cobalt supply to meet electric vehicle demand.[86]

Solar Energy Arrays and Wind Technologies

In the case of solar arrays and wind turbine technologies, USGS Minerals Information Center conducted a technical analysis of byproduct minerals that are contained in solar energy systems: silver, cadmium, tellurium, indium, gallium, selenium, germanium, and four of the REEs

---

[85] Battery types include LCO-C (lithium-cobalt-oxide), LMO-C (lithium manganese oxide), NCA-C (nickel cobalt aluminum), NMC-C (lithium nickel manganese cobalt), LFP-C (lithium iron phosphate), and LFP-LTO (lithium iron phosphate). All battery types use lithium and all except the LFP-LTO use graphite.

[86] Helbig 2017 reports on the study by Marscheider-Weidemann, et al., *Raw Materials for Emerging Technologies*, 2016.

used in wind technologies (dysprosium (Dy), neodymium (Nd), terbium (Te), and praseodymium (Pr)), using Clean Power Plan (CPP) and no-CPP scenarios.[87] USGS concluded that regardless of the scenario, the transition to renewables is very likely to accelerate in the coming decades and that a number of minor metals are likely to be constrained; thus rates of production of those metals would need to be increased to meet demand unless there are manufacturing shifts. The analysis concluded that the supply of heavy REEs used in permanent magnets (currently used in some of the new wind turbines) will not keep pace with demand from multiple end uses. The USGS assumed an aggressive electric vehicle market, the increased use of the magnets in electric vehicles, and new wind turbines' use of permanent magnets containing REEs. There is some disagreement over whether significant increases in REEs for magnets that would be used in wind energy systems will occur.[88]

Additionally, USGS concluded that the growth in demand for byproduct metals in solar and wind energy systems would compete with usage in electric and hybrid vehicles, and consumer electronics. The report asserts that a key uncertainty is net material intensity, i.e., the quantity of the byproduct metal required per unit of installed electric generating capacity, minus the amount of recycled material. For solar cells, net material intensity per generating capacity is dependent on the conversion efficiency of solar cells.

Related questions are: Where are the wind turbines and solar arrays being manufactured and which countries and firms would be impacted the most by any disruption in critical mineral supply for these end uses?

---

[87] Nassar, et al., *Byproduct Metal Requirements for U.S. Wind and Solar Photovoltaic Electricity Generation Up to 2040 Under Various Clean Power Plan Scenarios*, Applied Energy, 183 (2016) 1209-1226, http://dx.doi.org/10.1016/ j.apenergy.2016.08.062. The Clean Power Plan was an Obama Administration rule to cut carbon dioxide emissions by 32% of 2005 levels by 2030.

[88] Lovins, Amory, *Clean Energy and Rare Earths: Why Not to Worry*, May 21, 2017, http://thebulletin.org/2017/05/ clean-energy-and-rare-earths-why-not-to-worry. The articles states that magnet-free machines can perform any function required in electric vehicles and wind turbines and that the most effective substitute is better auto or turbine design.

Permanent Magnets

REEs in permanent magnets is another example of how materials analysis for end uses may inform understanding of critical minerals vulnerability. For example, some of the pertinent questions that might be raised with respect to permanent magnets include: How much Dy, Nd, Te, and Pr go into a neodymium-iron-boron (NdFeB) permanent magnet and what fraction of the total cost is each element? What are permanent magnet unit production costs and what portion of the total costs of a wind turbine or an automobile do the permanent magnets represent? And what is the likelihood and the economics of substitution?

Materials Review of Wind and Solar Energy Systems

Below are simplified examples of material requirements for wind and solar systems.

Materials for Wind Energy

Based on the Department of Energy Report, *20% Wind Energy by 2030*, wind power installations consist of four major parts: wind tower, rotor, electrical system, and drivetrain (e.g., generator, gearbox, and motor).[89] Most of the common large wind turbines have tower heights over 200 feet and rotor blades as long as 150 feet. The average rated capacity of an onshore wind turbine is between 2.5 megawatts (MW) and 3 MW.[90] DOE lists the following as the most important materials for large-scale manufacturing of wind turbines: steel, fiberglass, resins (for composites and adhesives), core materials, permanent magnets, and copper. Some aluminum and concrete is also required (see Table 6 below).

---

[89] U.S. Department of Energy, Energy Efficiency and Renewable Energy, *20% Wind Energy by 2030, Increasing Wind Energy's Contribution to U.S. Electricity Supply*, July 2008.

[90] Offshore wind farms are deploying much taller structures, with longer blades and greater MW capacity.

Table 6. Selected Materials for Wind Power

Turbine Materials	U.S. Supply	Comments
Permanent magnet	No U.S. production, little supply from secondary recovery	China produces nearly 75% of the world's permanent magnets. Significant production increases needed for future wind power needs.
Concrete	U.S. production	
Steel	U.S. production	
Aluminum	U.S. production (50% import reliant)	
Copper	U.S. production	
Glass fiber-reinforced plastic	U.S. production of fiberglass	
Carbon fiber-reinforced plastic	U.S. production	Globally, production of commercial grade carbon fiber is about 50 million lbs. per year. Significant production increases needed for future wind power needs.
Adhesives (petrochemical based)	U.S. production	
Core (petrochemical based)	U.S. production	
Battery technology containing the following materials		
Sodium sulfur	U.S. production	
Zinc-bromide	U.S. production	
Vanadium-redox	No U.S. production or reserves of vanadium	
Lithium-ion	Some U.S. production of lithium carbonate, small reserves	
Polysulfide-bromide	U.S. production	

Source: U.S. DOE, 20% Wind Energy by 2030 (2009) and Xcel 2007 Resource Plan, "Appendix E. Wind Storage Research and Experiments." Wilburn D.R., Wind Energy in the United States and Materials Required for the Land- Based Wind Turbine Industry From 2010 Through 2030. Scientific Investigations Report 2011-5036.

Notes: The critical minerals that could go into the manufacturing of wind turbines include the rare earth elements used in permanent magnets, vanadium and lithium for battery technology, and aluminum. These are shown in bold in the table.

DOE considers the raw materials for large-scale wind turbines to generally be in ample supply. Turbine manufacturing, however, would be 100% dependent on permanent magnet imports, primarily from China, as that country produces 75% of the world's permanent magnets which contain REEs (assuming certain drivetrains are used). But DOE and other wind power analysts also identify, as a potential concern, the need for increased manufacturing capacity for fiberglass and other components such as generators, and gear boxes. Wind power development trends at the time of the 20% Wind Energy by 2030 study were moving towards lighter-weight materials and high-strength composites such as glass fiber-reinforced plastic and carbon fiber-reinforced plastic. Increased production of fiberglass, commercial-grade carbon fiber, and permanent magnets (containing REEs) would be necessary if the United States were to achieve 20% wind energy by 2030. Recent analysis indicates that the offshore wind industry could be a major driver for increasing REE demand. There are indications that the larger turbines which are better suited for offshore locations, which also contain REEs, may be more reliable and require less maintenance than onshore turbines.[91]

Materials for Solar Energy

There are two major types of photovoltaic (PV) cells: crystalline silicon cells (most widely used) and thin film solar cells. The silicon based PV cells are combined into modules (containing about 40 cells) then mounted in an array of about 10 modules. Ethylene-vinyl acetate and glass sheets typically frame the PV module with additional aluminum frames for added protection.[92] Thin-film solar cells use layers of ultra-thin semi-conductor materials that can serve directly in rooftop shingles, roof tiles, and building facades. Thin-film PV cells have been noted to use cadmium-telluride or copper-indium-gallium-diselenide (see Table 7 below). A

[91] Fishman, Tomer, and T.E. Graedel, "Impact of the Establishment of U.S. Offshore Wind Power on Neodymium Flows," *Nature Sustainability*, vol. 2, April 2019; Dodd, Jan, "Rethinking the Use of Rare Earth Elements," *WindPower Monthly*, November 30, 2019, https://www.windpowermonthly.com/article/1519221/rethinkingtheuseofrare-earthelements.

[92] Vasilis M. Fthenakis, Hyung Chul Kim, and Erik Alsema, "Emissions from Photovoltaic Life Cycles," *Environment Science and Technology*, vol. 42, no. 6, 2008.

separate category of solar technology is concentrating solar power; these systems use mirrors to convert the sun's energy into heat and then into electricity.

Table 7. Selected Materials for Photovoltaic Solar Cells and Panels

Solar Energy Materials	U.S. Supply	Major Import Sources
Glass	Large sand production and reserves for making glass	Net exporter in 2008
Aluminum	U.S. supply	Net exporter in 2008
Copper	U.S. supply (32% import reliant)	Chile and Canada
Indium	Negligible U.S. supply from secondary sources	China, Japan, and Canada
Gallium	Negligible U.S. supply as byproduct of bauxite production	China, Ukraine and Germany
Tellurium	Byproduct of zinc production	Belgium, Canada, and China
Selenium	Byproduct of copper production	Belgium and Canada
Cadmium	Byproduct of copper production	Net exporter in 2008
Silicon Metal	Some U.S. production (less than 50% import reliant)	Brazil, South Africa, and Canada

Source: U.S. DOE, *Solar America Initiative*; "Emissions from Photovoltaic Cycles," *Environmental Science and Technology*, v. 2, no. 6, 2008.

Notes: The critical minerals that could go into the manufacturing of solar cells and panels include aluminum, indium, gallium, and tellurium: these are shown in bold.

SELECTED SUPPLY CHAIN ANALYSIS

With a supply chain analysis, it is just as important to know where new downstream capacity (processing, refining, and metals alloying) is being built or likely to be built in the world as it is to know the likely investors in upstream production capacity for critical minerals.

When looking at the complete supply picture it could be more easily determined where the potential risks are and what mitigation efforts may be available. Below, two illustrative supply chains are described: rare earth elements and tantalum.

Rare Earth Elements

REE Supply

Rare earth elements often occur with other elements, such as copper, gold, uranium, phosphates, and iron, and have often been produced as a byproduct. The lighter elements, such as lanthanum, cerium, praseodymium, and neodymium, are more abundant and concentrated and usually make up about 80%-99% of a total deposit. The heavier elements—gadolinium through lutetium and yttrium—are scarcer but very "desirable," according to USGS commodity analysts.[93]

Most REEs throughout the world are located in deposits of the minerals bastnaesite[94] and monazite.[95] Bastnaesite deposits in the United States and China account for the largest concentrations of REEs, while monazite deposits in Australia, South Africa, China, Brazil, Malaysia, and India account for the second-largest concentrations of REEs. Bastnaesite occurs as a primary mineral, while monazite is found in primary deposits of other ores and typically recovered as a byproduct. Over 90% of the world's economically recoverable rare earth elements are found in primary mineral deposits (e.g., in bastnaesite ores).[96]

REE Supply Chain

The supply chain for rare earth elements generally consists of mining, separation, refining, alloying, and manufacturing (devices and component parts). A major issue for REE development in the United States is the lack of refining, alloying, and fabricating capacity that could process any rare earth production.

An April 2010 GAO report illustrates the lack of U.S. presence in the REE global supply chain at each of the five stages of mining, separation,

---

93 DOI/USGS, *Rare Earth Elements-Critical Resources for High Technology*, Fact Sheet 087-02, 2006.

94 Bastnaesite is a mineral that may contain other rare earth elements.

95 Monazite is a mineral with the chemical composition of (Ce, La, Nd, Th) PO4 also may contain other rare earth elements.

96 DOI/USGS *International Strategic Minerals Inventory Summary Report—Rare Earth Oxides*, by Wayne Jackson and Grey Christiansen, Circular 930 N, 1993.

refining oxides into metal, fabrication of alloys, and the manufacturing of magnets and other components. According to the 2010 GAO report, China produced about 95% of the REE raw materials and about 97% of rare earth oxides, and was the only exporter of commercial quantities of rare earth metals (Japan produced some metal for its own use for alloys and magnet production). About 90% of the metal alloys were produced in China, and China manufactures 75% of the NdFeB magnets and 60% of the samarium cobalt (SmCo) magnets. Thus, even as U.S. rare earth production ramps up, without significant supply chain investments, much of the processing and metal fabrication would likely occur in China.

In the case of rare earths, it is not enough to develop REE mining operations outside of China alone without building the value-added refining, metal production, and alloying capacity that would be needed to manufacture component parts for end-use products. According to rare earth analyst Jack Lifton, vertically integrated companies may be more desirable. It may be the best way to secure investor financing for REE production projects.[97] Joint ventures, consortiums, and cooperatives could be formed to support production at various stages of the supply chain at optimal locations around the world. Each investor or producer could have equity and offtake commitments. Where U.S. firms and U.S. allies invest may contribute to meeting the goal of providing a secure and stable supply of REEs, intermediate products, and component parts needed for the assembly of end-use products.

In 2019, rare earth analyst James Kennedy of ThREE Consulting writes that China's dominance and "absolute advantage" in the rare earth space is fundamentally reflected in its R&D efforts at its national labs and the Baotou Research Institute of Rare Earths in the fields of basic sciences, materials science, and rare earth metallurgy.[98] ThREE Consulting has shown that China has filed more rare earth patents than the rest of the

---

[97] The Gold Report, *Rare Earth Strategic Supplies More Important Than Price,* Industrial Metals/Minerals Interview with Jack Lifton, December 14, 2009.

[98] Kennedy, James, "China Solidifies Dominance in Rare Earth Processing," *National Defense Magazine*, March 21, 2019, https://www.nationaldefensemagazine.org/articles/2019/3/21/viewpoint-china-solidifies-dominance.

world combined and Kennedy states that patents acquired in the rare earth space are likely a proxy for next generation rare earth-related technology.

China's whole-of-government approach in the field of rare earths and other critical minerals may keep China in its position of dominance for the foreseeable future.

Tantalum

Tantalum is a metallic element contained in the mineral tantalite and is extracted from primary and placer mineral deposits.[99] It often occurs with niobium but is also present with other minerals such as rare earths, uranium, and cassiterite (tin ore). Tantalum has been produced as a primary product, a co-product, and as a byproduct of other ores. Tantalum's high melting point (3,000 degrees Centigrade) and corrosion resistance makes it super-capacitive, (i.e., characterized by a high capacity to store and release electrical charges). This metal, which is used in numerous high- tech electronic devices, is produced and traded in conflict areas in Central Africa; thus, in certain instances, tantalum is classified as a conflict mineral and subject to disclosure rules promulgated from the Dodd-Frank Wall Street Reform and Consumer Protection Act (P.L. 111-203, 15 U.S.C. §78).[100] Section 1502 of the law includes a sense of the Congress that conflict minerals in the Democratic Republic of the Congo or adjoining countries are financing extreme levels of violence in the DRC.

Tantalum Supply

There are four major sources of tantalum market supplies: primary production (industrial and artisanal[101]); tin slag processing; scrap

---

[99] Placer deposits are those formed when material is removed from primary deposits and accumulates in other locations, typically after being moved by water (alluvial sedimentation), other forms of erosion, or other natural forces.

[100] Section 1502, which amends the Securities and Exchange Commission Act (SEC) of 1934, defines conflict minerals as "columbite-tantalite (coltan), cassiterite, gold, wolframite, or their derivatives." Coltan is a colloquial expression for the combined columbite (niobium)–tantalite ores that are found in Central Africa.

[101] Artisanal mining refers to small-scale operators, usually very labor intensive (using picks and shovels), as compared with capital intensive industrial production.

reprocessing and recycling; and byproduct production (also referred to as secondary concentrate).[102] Primary production accounts for about 70% of global supply. Historically, tantalum obtained from tin slag (waste) was primarily produced in Malaysia, Thailand, and Brazil. Tantalum has also been a byproduct of niobium, titanium, tin, and uranium produced in Malaysia, Brazil, China, and Russia.

Recycled tantalum contributes to 30% of global supply, mostly recovered from "pre-consumer scrap" at the manufacturing plant. The United States and Mexico account for 61% of tantalum scrap recovery and it is estimated that scrap could provide 50% of global tantalum supply by 2025.[103]

Based on USGS data, Brazil, Canada, Mozambique, and Nigeria were countries that led in primary tantalum production during the 1970s. Brazil and Canada continued to be the major producing countries in the 1980s.[104] Australia took over the top spot in the late 1980s and 1990s, followed by Brazil until 2009, after which no primary production was reported for Australia by the USGS. The Australian mines were closed following the 2008 recession, reopened in 2012, but closed again shortly thereafter in 2012. Since about 2009, it has been noted by several sources that the DRC, with tens of thousands of artisanal miners, is a leading producing country (see Table 4).[105] Recorded production for tantalum by the USGS indicates a shift in production—at least what has been reported—since 2000 from Australia and Brazil, to the DRC and Rwanda.[106]

Over the past several decades, there were material gaps in the publically available data for tantalum; production data reported has been much less than processor receipts. In one example, the average producer's supply to total processor's receipts gap measured over six quarters was

---

[102] Ulric Schwela, "Focus: Tantalum: Regulating the Supply Chain," *Mining Journal*, March 22, 2013, https://www.mining-journal.com/africa/news/1164441/focus-tantalum-regulating-supply-chain.

[103] N.A. Mancheri, et al., "Resilience in the Tantalum Supply Chain," *Resources, Conservation, and Recycling*, October 18, 2017.

[104] USGS, *Shift in Global Tantalum Production, 2000-2014*, Fact Sheet 2015-3079, December 2015.

[105] USGS, Mineral Commodity Summaries, Various Years.

[106] USGS, *Shift in Global Tantalum Production, 2000-2014*.

73%. On average, reported production represents about 27% of total processors' receipts over the period. There was an average material difference of 381 metric tons.[107]

Part of the explanation for such reporting patterns may be the highly unregulated nature of tantalum ore production and trade in Central Africa.[108] High production in the unreported (informal) sector of the mining community drove prices down and forced many of the major production regions to close their operations. With low prices, investor interest is limited; investors are thus constrained by high risk in greenfield projects, (i.e., new projects or work that does not follow previous work).[109]

The USGS data does not reflect the amount of production from unauthorized (often illegal) mining operations—usually artisanal mining operations. The USGS collects its data from a variety of sources but considers the tantalum industry as operating under "a shroud of secrecy" with incomplete access to data and not very transparent. Generally, there is insufficient data to make definitive determinations on the true production, capacity, and reserve levels for tantalum on a global basis. There are several reasons for this supply/demand material difference, including the following:[110]

- Nonreporting or under-reporting all forms of supply (primary, byproduct, tin slag, and scrap) through the Tantalum-Niobium International Study Center (TIC) or elsewhere.

---

[107] Data from USGS Mineral Commodity Receipts and the Titanium-Niobium International Study Center (TIC) various years. The TIC is an international trade association comprising of around 85 members, all involved in the industries of tantalum and/ or niobium, at various positions along the supply chain.

[108] Raimund Bleischwitz, et al., "Coltan from Central Africa, International Trade and Implications for Any Certification," *Resources Policy*, March 2012.

[109] N.A. Mancheri, et al., "Resilience in the Tantalum Supply Chain," *Resources, Conservation, and Recycling*, October 18, 2017.

[110] Raimund Bleischwitz, et al., "Coltan from Central Africa, International Trade and Implications for Any Certification," *Resources Policy*, March 2012. United Nations sanctions monitors also track trade in coltan ore in Central Africa and have repeatedly documented unofficial, often black market cross-border movements of the ore. Also see, USGS Fact Sheet 2015-3079, *Shift in Global Tantalum Mine Production, 2000-2014*, December 2015. Mancheri, et al. reports on this information gap as well in "Resilience in the Tantalum Supply Chain," *Resources, Conservation, and Recycling*, October 18, 2017.

- High inventories. Several analysts have noted that since the recession of 2008 many companies were selling from their above-ground stocks.
- Illicit mining and trading. There are well-established networks for smuggling tantalum and other minerals out of Central Africa (and elsewhere) and into the marketplace.

Dependence on Africa's supply and that disruption could have consequences, e.g., price rises. Africa provides 80% of the primary tantalum production (60% from the DRC and Rwanda) as China dominates downstream processing and manufacturing capacity. The illicit mining component in the tantalum market makes it vulnerable and possibly unsustainable because it prevents large-scale producers from entering the market. Illegal tantalum trade has long-term implications for the entire supply chain leading to lower investment in all phases of the supply chain.[111]

In 2016, the USGS listed Australia and Brazil as having 85% of the world's tantalum reserves, but the USGS regularly states that data is not available for other countries or is just unknown. The USGS lists Australia, Brazil, and Canada as having the majority of the world identified tantalum resources.

The Tantalum Supply Chain

In 2017, Mancheri, et al., published a study that assessed the tantalum supply chain for regional production dependence, the potential for supply disruptions, and mechanisms to prevent disruptions using a "resiliency" of supply model.[112] This method examines four resilience of supply indicators: diversity of supply, material substitution, recycling, and stockpiling, and is dependent on three factors: resistance, rapidity, and

---

[111] NA Mancheri, et al., "Resilience in the Tantalum Supply Chain," *Resources, Conservation, and Recycling*, October 18, 2017.

[112] According to Mancheri, "resilience theory" shows how a supply chain would respond to disruptions in short-term and long term-constraints. This theory is discussed in Mancheri, et al., "Resilience in the Tantalum Supply Chain," *Resources, Conservation, and Recycling*, October 18, 2017.

flexibility. Mancheri's study concludes that the tantalum market is flexible and resilient based on its handling of unreported and presumably illegal trade along with its impact on conventional large-scale tantalum producers. Mancheri's study concluded that stockpiling and substitution can mitigate some supply disruption.

Generally, tantalum follows the following supply chain steps:[113]

- The primary ore is crushed and milled into an ore concentrate which is further refined into oxides (metal or powder) or K-Salt (which is reduced to tantalum metal),[114] which is used for the manufacture of capacitors, wire, super alloys, and other fabricated forms. Downstream manufacturers use these materials for parts that are used by consumer product manufacturers and others.[115] China has 16 tantalum processing plants; the United States has one, according to the Mancheri study. There are four processing plants in Germany and four in Japan.
- The metal or powder form is then used by electronics manufacturers to produce capacitors and other products. The manufactured parts are shipped to consumer product producers such as Motorola, Sony, Apple, Dell, and others. China dominates the production of capacitors.

CURRENT POLICY FRAMEWORK

U.S. Mineral Policy

As noted in two key statutes, the current goal of U.S. mineral policy is to promote an adequate, stable, and reliable supply of materials for U.S.

---

[113] British Geological Survey, *Niobium-Tantalum*, April 2011, pp. 11-13.

[114] Tantalum (Ta) extraction uses solvent extraction ammonium to precipitate tantalum as Ta-hydroxide which is calcined to form Ta-oxide or potassium tantalum fluoride (K-Salt). K-Salt is directly crystallized by adding potassium fluoride to the extract solution. The molten sodium is used to form Ta metal.

[115] British Geological Survey, *Niobium-Tantalum*, April 2011, pp. 11-13.

national security, economic well-being, and industrial production. U.S. mineral policy emphasizes developing domestic supplies of critical materials and encourages the domestic private sector to produce and process those materials.[116] But some raw materials do not exist in economic quantities in the United States, and processing, manufacturing, and other downstream ventures in the United States may not be cost competitive with facilities in other regions of the world. However, there have been public policies enacted or executive branch measures taken (for example, the percentage depletion allowance[117] for U.S. mining operations and royalty-free production on public domain lands) to offset the U.S. disadvantage of its potentially higher-cost operations. The private sector also may achieve lower- cost operations with technology breakthroughs.

Based on this policy framework, Congress has held numerous legislative hearings on the impact of the U.S. economy's high import reliance on many critical materials, and on a range of potential federal investments that would support the development of increased domestic production and production from reliable suppliers. There has been a long-term policy interest in mineral import reliance and its impact on national security and the U.S. economy.

General Mining Law of 1872: Mining on Federal Lands

Mining of locatable minerals (also referred to as hardrock minerals) on federal lands is governed primarily by the General Mining Law of 1872 (30 U.S.C. §§21-54). The original purposes of the Mining Law were to promote mineral exploration and development on federal lands in the

---

[116] U.S. mineral policies provide a framework for the development of domestic metal mineral resources and for securing supplies from foreign sources. Specifically, the Mining and Minerals Policy Act of 1970 (30 U.S.C. §21a) declared that it is in the national interest of the United States to foster the development of the domestic mining industry "... including the use of recycling and scrap." The National Materials and Minerals Policy, Research and Development Act of 1980 (30 U.S.C. §1601) declares, among other things, that it is the continuing policy of the United States to promote an adequate and stable supply of materials necessary to maintain national security, economic well-being, and industrial production, with appropriate attention to a long-term balance between resource production, energy use, a healthy environment, natural resources conservation, and social needs. There is also a provision to develop an early warning system for critical materials.

[117] A percentage depletion allowance is a tax deduction to recover investments in mineral properties, (e.g., property acquisition costs and capitalized exploration expenses).

western United States, offer an opportunity to obtain a clear title to mines already being worked, and help settle the West. The Mining Law grants free access to individuals and corporations to prospect for minerals on open public domain lands, and allows them, upon making a discovery, to stake (or "locate") a claim on the deposit. A valid claim entitles the holder to develop the minerals. The 1872 Mining Law originally applied to all valuable mineral deposits except coal (17 Stat. 91, 1872, as amended).

Public domain lands are those retained under federal ownership since their original acquisition by treaty, cession, or purchase as part of the general territory of the United States, including lands that passed out of but reverted back to federal ownership. "Acquired" lands—those obtained from a state or a private owner through purchase, gift, or condemnation for particular federal purposes rather than as general territory of the United States—are subject to leasing only and are not covered by the 1872 Law. Acquired lands are governed under the authority of the Mineral Leasing for Acquired Lands Act of 1947.

Under the General Mining Law, mineral claims may be held indefinitely without any mineral production. Once lands were patented to convey full title to the claimant, the owner could use the lands for a variety of purposes, including nonmineral ones. However, using land under an unpatented mining claim for anything but mineral and associated purposes violates the General Mining Law. Critics believe that many claims are held for speculative purposes. However, industry officials argue that a claim may lie idle until market conditions make it profitable to develop the mineral deposit. Congress has placed a moratorium on patenting lands since 1994 under annual appropriation bills.[118]

The vast majority of mineral production in the United States occurs on private land and is regulated by the states which may use a leasing and permitting framework. The regulatory framework described below applies primarily to minerals produced on federal land but has implications for the entire U.S. mining industry.

---

[118] For example, see P.L. 109-54, Section 408 for standard appropriation language prohibiting further patenting of mining claims.

There is debate over whether streamlining the permitting process on federal lands would make investing in mining in the United States more attractive or would incentivize investors.

Proponents of streamlining the framework maintain that mining firms would be more likely to invest in the United States given a more rapid turnaround of the mine permitting process.

However, mining firms have multi-factor decision making processes; they go to where the minerals are, and they often look for low political and country risk (good governance) and a sense of certainty of the regulatory environment, as well as low-cost production opportunities.

A debate has emerged over the past several decades over whether the federal government should impose a royalty on the value of minerals produced on public lands, as is the practice on other lands in the United States (i.e., state lands and private lands) and other parts of the world. Further discussion of this debate is beyond the scope of this chapter.

Federal Land Management and Mineral Development: Regulatory Framework for Mineral Development on Federal Land

Mineral development activities in the United States are subject to a suite of federal regulatory requirements. The specific statutes and regulations that will apply and how compliance is accomplished will vary depending on the specific mineral development project (e.g., specific actions may be required for compliance with federal law if the mining project may affect a federally protected species). That is, for mining on federal lands, there are various federal regulatory requirements that may apply in addition to the Federal Mining Law of 1872. These requirements encompass environmental reviews, adequate proof of financing, permits, surface management requirements, bonding, and public participation, among other requirements. The Appendix provides a list of the selected statutes and regulations related to mineral development on federal land. A discussion of the regulatory compliance process and the various federal, state, and other entities that may be involved is beyond the scope of this chapter. The following discussion focuses on the regulatory framework associated with management of and access to minerals for development on federal land.

During the 1960s and 1970s, the Multiple Use Sustained Yield Act (16 U.S.C. §§528-531), Wilderness Act of 1964 (16 U.S.C. §§1131-1136), National Forest Management Act of 1976 (43 U.S.C. §§1701 et seq.), National Environmental Policy Act of 1969 (NEPA, 42 U.S.C. §§4321 et seq.), and Federal Land Policy Management Act (FLPMA) (43 U.S.C. §1701 et seq.) addressed environmental protection, multiple use, and management of federal land generally. By imposing requirements on agency actions, these acts have affected mineral development under both the leasing system and the General Mining Law of 1872 claim-patent system. The General Mining Law contains no direct environmental controls, but mining claims are subject to all general environmental laws as a precondition for development.

The Bureau of Land Management (BLM)[119] administers the mineral program on all federal land but other land managing agencies, such as the Forest Service (FS) must approve surface disturbing activity on its land. BLM and FS use the mine plan review process (which includes mining methods and reclamation plans) to determine the validity of the mine proposal and to determine how extensive of an environmental review is required under the Federal Land Policy and Management Act of 1976.

Federal Land Policy Management Act

Under the Federal Land Policy and Management Act of 1976, Resource Management Plans (RMPs) are required for tracts or areas of

---

[119] The Bureau of Land Management is an Interior Department agency that is responsible for approximately 700 million acres of federal subsurface minerals, and supervises the mineral operations on about 56 million acres of Indian trust lands. According to the BLM, approximately 150 million acres have been withdrawn from mineral entry, leasing, and sale, subject to valid existing rights. Lands in the National Park System (except National Recreation Areas) and the Wilderness Preservation System are among those that are statutorily withdrawn. Also, of the 700 million acres, mineral development on 182 million acres is subject to the approval of the surface management agency (e.g., the Forest Service), and must not be in conflict with land designations and plans.

public lands prior to development. BLM must consider environmental impacts during land-use planning when RMPs are developed and implemented.

RMPs can cover large areas, often hundreds of thousands of acres across multiple counties.[120]

Through the land-use planning process, BLM determines which lands are open for mining claims and potential development.[121]

Regarding land use plans FLPMA states: "the Secretary [of the Interior] shall with public involvement and consistent with the terms and conditions of this Act, develop, maintain and, when appropriate, revise land use plans which provide by tracts or areas for the use of the public lands."[122] Current planning regulations require preparation of an environmental review document for the land use plans under the National Environmental Policy Act.[123]

FLPMA requires that RMPs reflect diverse uses—such as timber, grazing, wildlife conservation, recreation, and energy—and consider the needs of present and future generations. Impacts of various uses are identified early in the process so that they can be weighed equitably against one another by the BLM. The plans are also intended to weigh the various benefits associated with public lands.

Withdrawals from Mineral Entry and Access to Federal Land

The President and executive branch agencies historically issued executive orders, secretarial orders, and public land orders to withdraw federal lands from mineral entry and other uses under what was viewed as the President's authority, including certain statutory authorities such as the Antiquities Act (34 Stat. 225). Since 1976 executive withdrawals are governed by FLPMA. FLPMA repealed earlier land withdrawal authorities. Withdrawals of parcels exceeding 5,000 acres require congressional approval.[124]

---

[120] 43 U.S.C. §1712.
[121] Ibid.
[122] Ibid.
[123] 43 C.F.R. 1601.06.
[124] Congress may adopt a concurrent resolution not approving the withdrawal.

A withdrawal pursuant to FLPMA restricts the use of land under the multiple-use management framework, typically segregating the land from some or all public land laws as well as some or all of the mining and mineral leasing laws for a period of 20 years.[125] Initially, the area is segregated for two years during which time an environmental review is conducted to determine whether a longer-term withdrawal of 20 years is warranted. The longer-term withdrawal is often subject to renewal by the Department of the Interior.

The withdrawal can be temporary or permanent. Under this section of the code the Secretary of the Interior may make, modify, extend, or revoke withdrawals.[126]

Generally, federal land withdrawals are subject to valid existing rights, meaning that the minerals rights holder may develop those minerals subject to terms of the federal land-managing agency (e.g., the National Park Service, BLM, or the Forest Service).

Mineral industry representatives maintain that federal withdrawals inhibit mineral exploration and limit the reserve base even when conditions are favorable for production. Thus, they state that without new reserves or technological advancements mineral production costs may rise. They further contend that higher domestic costs may lead to greater exploration on foreign soil, potentially boosting U.S. import dependence.

Critics of U.S. mineral development state that mining often is an exclusive use of land inasmuch as it can preclude other uses, and that in many cases there is no way to protect other land values and uses short of withdrawal of lands from development under the General Mining Law. They point to unreclaimed areas associated with previous hardrock mineral development, Superfund sites related to past mining and smelting, and instances where development of mineral resources could adversely affect or destroy scenic, historic, cultural, and other resources on public land.

---

[125] Under FLPMA, (43 U.S.C.A. §1702(j)), Congress defined a withdrawal as "withholding an area of Federal land from settlement, sale, location or entry under some or all of the general land laws for the purpose of limiting activities under those laws in order to maintain other public values in the area or reserving the area for a particular public purpose or program; or transferring jurisdiction over an area of federal land from one department, bureau or agency to another department, bureau or agency."

[126] Ibid.

Congressional debate has been ongoing for decades over how much federal land should be available for the extractive industries or other uses and how much should be set aside (e.g., off limits or restricted) for conservation or environmental purposes.

SELECTED CRITICAL MINERALS-RELATED LEGISLATION IN THE 115TH AND 116TH CONGRESSES

116th Congress

H.R. 2531, National Strategic and Critical Minerals Production Act, introduced by Representative Mark E. Amodei on May 7, 2019, and referred to House Committee on Natural Resources. The bill would define critical and strategic minerals and seeks to streamline the federal permitting process for domestic mineral exploration and development. It would establish responsibilities of the "lead" federal agency to set mine permitting goals, minimize delays, and follow time schedules when evaluating a mine plan of operations. The review process would be limited to 30 months, and the bill would establish the priority of the lead agency maximizing the development of the mineral resource while mitigating environmental impacts.

H.R. 2500, National Defense Authorization Act (NDAA) for Fiscal Year 2020, reported in the House. The bill would require the Secretary of Defense to provide guidance on acquiring items containing rare earth elements and guidance on establishing a secure rare earth materials supply chain within the United States. The bill provides authority for the Secretary to acquire rare earth cerium and lanthanum compounds and electrolytic manganese metal. And further, for DOD purposes, the bill would prohibit the acquisition of tantalum from nonallied foreign nations.

The reported Senate version (S. 1790) of the FY2020 NDAA does not contain similar language.

S. 1317, American Mineral Security Act, introduced by Senator Murkowski on May 2, 2019, and referred to the Senate Committee on Energy and Natural Resources.

The bill would define what critical minerals are, but also would request that the Secretary of the Interior establish a methodology that would identify which minerals qualify as critical. The Secretary of the Interior would be required to maintain a list of critical minerals. The bill would establish an analytical and forecasting capability on mineral/metal market dynamics as part of U.S. mineral policy. The Secretary of the Interior would be required to direct a comprehensive resource assessment of critical mineral resource potential in the United States, assessing the most critical minerals first.

The bill would require that an agency review and report be intended to facilitate a more efficient process for critical minerals exploration on federal lands, and specifically would require performance metrics for permitting mineral development activity and report on the timeline of each phase of the process.

The bill would require that the Department of Energy establish an R&D program to examine the alternatives to critical minerals and explore recycling and material efficiencies through the supply chain. The Department of the Interior would be required to produce an Annual Critical Minerals Outlook report that would provide forecasts of domestic supply, demand, and price for up to 10 years.

The Secretary of Labor, in consultation with the National Science Foundation and other relevant institutions, would be required to assess the availability of domestic technically trained personnel in the exploration production, manufacturing, recycling, forecasting, and analysis of minerals critical to the United States, noting, among other things, skills in short supply now, and those projected to be in short supply in the future. The Secretary would be required to design an interdisciplinary curriculum study on critical minerals and further, establish a competitive grants program for new faculty positions, internships, equipment needs, and research related to critical minerals. There would be $50 million authorized to carry out this act each year for fiscal years 2020-2029.

115th Congress

H.R. 520, National Strategic and Critical Minerals Production Act, introduced by Representative Mark E. Amodei on January 13, 2017, and referred to House Committee on Natural Resources. This bill is similar to H.R. 2531 described above (in the 116th Congress).

H.R. 1407, METALS Act, introduced by Representative Duncan Hunter on March 7, 2017, and referred to the House Committee on Armed Services.

This bill would have established a strategic materials investment fund and allowed the Secretary of Defense to provide loans for domestic production and domestic processing of strategic and critical materials, and supported the development of new technologies for more efficient processing of strategic and critical materials.

For fiscal years 2018 through 2023, 1/10 of 1% of the amounts appropriated for "covered programs" would have been deposited into the fund. Covered programs would have been all major defense acquisition programs for development or procurement of aircraft or missiles. The bill would have established a prohibition on sale of domestic rare earth mines to foreign firms.

H.R. 5515 (P.L. 115-232), John S. McCain National Defense Authorization Act for Fiscal Year 2019, included a provision to direct the Secretary of Defense to purchase rare earth permanent magnets and certain tungsten, tantalum, and molybdenum from sources outside of China, Russia, North Korea, and Iran to the extent possible.

S. 1460, Energy and Natural Resources Act of 2017, Subtitle D—Critical Minerals, introduced by Senator Murkowski on June 18, 2017, and referred to the Senate Committee on Energy and Natural Resources. This bill is similar to S. 1317 above (in the 116th Congress).

S. 145, National Strategic and Critical Minerals Production Act (similar to H.R. 520 in the 115th Congress), introduced by Senator Heller on January 12, 2017, and referred to the Senate Committee on Energy and Natural Resources.

Previous Congresses

Similar bills on critical minerals were introduced in earlier Congresses. For example, in the 113th Congress, there was S. 1600, Critical Minerals Policy Act of 2013, and H.R. 761, the National Strategic and Critical Minerals Production Act of 2013, which passed the House on September 18, 2013. Another bill in the 113th Congress, H.R. 4883, the National Rare Earth Cooperative Act of 2014, proposed to advance domestic refining of heavy rare earth oxides and the safe storage of thorium for future uses using a cooperative ownership approach. Thorium is associated with certain rare earth deposits and waste materials. The cooperative would have operated under a federal charter composed of suppliers and consumers as owners.

Additional Policy Options

This section provides a discussion of selected policy options related to critical minerals that were included in legislation introduced in the 115th and 116th Congresses. In addition to weighing the advantages and disadvantages of the various policy options discussed above and below, policymakers have the option of maintaining the status quo of current policies.

Minerals Information Administration

The USGS could establish a Minerals Information Administration for information and analysis on the global mineral/metal supply and demand picture. Companies producing minerals on public lands could be required to report production data to the federal agency.

Greater Exploration for Critical Minerals

Encouragement of greater exploration for critical minerals in the United States, Australia, Africa, and Canada could be part of a broad international strategy. There are only a few companies in the world that can provide the exploration and development skills and technology for critical mineral development. These few companies are located primarily in the above four regions and China, and may form joint ventures or other types of alliances for R&D, and for exploration and development of critical mineral deposits worldwide, including those in the United States.

Whether there should be restrictions on these cooperative efforts in the United States is a question for congressional deliberations.

Other Policy Options

Other action by Congress could include oversight of free trade issues associated with critical mineral supply. Two raw material issues associated with China export restrictions were taken up by the World Trade Organization (WTO). One case, settled in 2011, was filed by the United States against China and was related to restrictions on bauxite, magnesium, manganese, silicon metal, and zinc, among others (using export quotas and export taxes). The other case, resolved in 2012, was filed by the United States, Japan, and the European Union on export restrictions of rare earth oxides, tungsten, and molybdenum. The WTO ruled against China in both cases, concluding that China did not show the link between conservation of resources or environmental protection (and protection of public health) and the need for export restrictions.

The United States could support more trade missions; support U.S. commercial delegations to China and other mineral-producing countries; and assist smaller and less-developed countries in improving their governance capacity. Although there are concerns that trade tariffs with China could impact the prices and availability of critical minerals and downstream metals imported from China, the effects would depend on the specifics of the tariffs as well as the particular mineral and metal involved.

Additional Considerations

In China and other emerging economies, economic development will continue to have a major impact on the world supply and availability of raw materials and downstream products. Various countries may be faced with making adjustments to secure needed raw materials, metals, and finished goods for national security and economic development. China, Japan, and others are already actively engaged in securing reliable mineral supplies. Many firms have moved to China to gain access to its market, raw materials, or intermediate products, and generally lower-cost minerals production. At the same time, China is seeking technology transfer from many of these firms to expand its downstream manufacturing capacity. Despite China's current overcapacity and increased exports of some commodities, in the long run it may be in China's interest to use its minerals (plus imports) for domestic manufacturing of higher-valued downstream products (e.g., component parts and consumer electronics). Higher-cost, inefficient facilities and mines may close, resulting in China seeking more imports as mining industry consolidations are implemented.

The effects on China's dominance in the supply and demand of global raw materials could be addressed in part through consistent development of alternate sources of supply, use of alternative materials when possible, efficiency gains, aggressive R&D in development of new technologies, and comprehensive minerals information to support this effort. China is likely entering an era of fewer raw material exports which may instigate long-term planning by the private sector and government entities that want to meet U.S. national security, economic, and energy policy interests and challenges. Some stakeholders may seek to have some concerns addressed through the WTO.

Additional questions that may be deliberated by Congress include how long would it take to develop the skill set in the United States for downstream manufacturing activities? Would an international educational exchange program with those countries already involved in the refining and recycling of critical minerals be appropriate?

More analysis would be useful to investigate U.S. firms' capacity to adjust to supply bottlenecks such as restrictions in other countries' exports, underinvestment in capacity, materials use in other countries and domestically, single source issues, strikes, power outages, natural disasters, political risk, and lack of substitutes. Having such analysis and understanding may inform public policy.

More information could inform deliberations as Congress and other policymakers evaluate the available policy options and their effectiveness at minimizing the risk of potential supply interruption of critical and strategic minerals and metals.

APPENDIX. SELECTED STATUTES AND REGULATIONS RELATED TO MINING ON FEDERAL LANDS

Selected Statutes that May Impact Mining Activities on Federal Lands (in alphabetical order)

- American Indian Religious Freedom Act (P.L. 95-341)
- Clean Air Act, 42 U.S.C. §7401 et seq.
- Clean Water Act, 33 U.S.C. §1251 et seq.
- Endangered Species Act, 16 U.S.C. §1531 et seq.
- Federal Land Policy and Management Act, 43 U.S.C. §1701-1784.
- Federal Mine Safety and Health Act of 1977 (P.L. 95-164).
- General Mining Law of 1872, 30 U.S.C. §21-54.
- Historic Preservation Act (P.L. 89-665).
- Mineral Leasing For Acquired Lands Act of 1947, 30 U.S.C. §351-359
- Mining and Minerals Policy Act of 1970, 30 U.S.C. §21a.
- National Environmental Policy Act, 42 U.S.C. §4321 et seq.

- National Forest Management Act 16 U.S.C. §1600-1687.
- National Materials and Minerals Policy, Research, and Development Act of 1980, 30 U.S.C. §1601.
- Resource Conservation and Recovery Act, 42 U.S.C. §6901 et seq.
- Toxic Substance Control Act (P.L. 94-469).

Mining-Specific Regulations

- Bureau of Land Management (BLM): 43 C.F.R. 3809—Regulations on surface management
- U.S. Forest Service (FS): 36 C.F.R. Part 228—Regulations on minerals

In: Key Congressional Reports for June 2019 ISBN: 978-1-53616-662-0
Editor: Stephanie A. Betts

Chapter 6

THE NATIONAL TRAILS SYSTEM: A BRIEF OVERVIEW*

Mark K. DeSantis and Sandra L. Johnson

SUMMARY

The National Trails System was created in 1968 by the National Trails System Act (16 U.S.C. §§1241-1251). The system includes four types of trails: (1) national scenic trails (NSTs), which display significant physical characteristics of U.S. regions; (2) national historic trails (NHTs), which follow travel routes of national historical significance; (3) national recreation trails (NRTs), which provide outdoor recreation accessible to urban areas; and (4) connecting or side trails, which provide access to the other types of trails. As defined in the act, NSTs and NHTs are long-distance trails designated by acts of Congress. NRTs and connecting and side trails may be designated by the Secretaries of the Interior and Agriculture with the consent of the federal agency, state, or political subdivision with jurisdiction over the lands involved.

* This is an edited, reformatted and augmented version of Congressional Research Service Publication No. R43868, Updated June 10, 2019.

Congress plays an ongoing role in shaping the National Trails System through legislation and oversight. Broad issues for Congress include, among others:

- whether and where to establish new trails in the system,
- whether to establish new trail categories (such as "national discovery trails"), and
- how much funding to provide to agencies for trail management.

When designating individual trails, Congress has considered issues such as:

- how to balance trail designation with other potential land uses,
- how to address federal land acquisition, and
- whether to make specific provisions for trail use that may differ from those authorized in the overall act.

Congress has established 11 NSTs and 19 NHTs, as well as several NRTs (although recreation trails are more typically designated administratively). In addition, the Secretaries of the Interior and Agriculture have designated just under 1,300 NRTs and seven connecting or side trails. The scenic, historic, and connecting trails are federally administered by the National Park Service (NPS) and/or the Bureau of Land Management (BLM) in the Department of the Interior, or the U.S. Forest Service (FS) in the Department of Agriculture, with cooperation from states and other entities to operate nonfederal trail segments. The roughly 1,300 national recreation trails are typically managed by states, localities, and private organizations, except where they cross federal lands. The act limits federal land acquisition for the trails system, with specific provisions for different trail types.

Each federal agency with management authority over national trails has its own budget for trail administration and management. Trails have also received funding from federal transportation programs, private donations, permits and fees, and local excise taxes, among other sources.

Uses of the national trails may include, but are not limited to, bicycling, cross-country skiing, day hiking, equestrian activities, jogging or similar fitness activities, overnight and long-distance backpacking, snowmobiling, and surface water and underwater activities. Provisions for motorized vehicle use vary among the different types of trails.

In March 2019, the 116th Congress passed and the President signed the John Dingell, Jr. Conservation, Management, and Recreation Act (P.L. 116-9), which extended existing national scenic and historic trails and directed the study of an additional trail route for potential addition to the system. Other legislation introduced in the 116th Congress would designate a new national historic trail (H.R. 434) and establish a new tax credit for conservation contributions to NSTs (S. 809 and H.R. 1727). As in earlier Congresses, a bill (H.R. 726) has also been introduced to add a new type of trail—national discovery trails—to the system.

The National Trails System Act of 1968 (16 U.S.C. §§1241-1251)[1] established the Appalachian and Pacific Crest National Scenic Trails, and authorized a national system of trails to provide outdoor recreational opportunities and to promote access to the nation's outdoor areas and historic resources. Since the act's passage, the system has grown to encompass trails in every U.S. state, the District of Columbia, and Puerto Rico.

The system has expanded over time and now includes four types of trails:

- National Scenic Trails (NSTs) display significant characteristics of the nation's "physiographic regions,"[2] representing desert, marsh, grassland, mountain, canyon, river, forest, or other areas. NSTs provide for outdoor recreation and for the conservation and enjoyment of significant scenic, historic, natural, or cultural qualities.
- National Historic Trails (NHTs)[3] identify and protect travel routes of national historic significance, along with associated remnants and artifacts, for public use and enjoyment. NHTs can include land or water segments,[4] marked highways paralleling the route, and sites that together form a chain or network along the historic route.

---

[1] P.L. 90-543, as amended.

[2] Descriptions of the trail types are from 16 U.S.C. §1242.

[3] The National Parks and Recreation Act of 1978 (P.L. 95-625) amended the original act to establish national historic trails (NHTs) as a type of trail within the system, and it established the first four NHTs.

[4] The 109th Congress established the Captain John Smith Chesapeake National Historic Trail (P.L. 109-418), the nation's first all-water national historic trail. The Department of the Interior (DOI) has also established a National Water Trails System as a class of national recreation trails. For more information, see National Park Service, *Budget Justifications and Performance Information, Fiscal Year 2020*, p. ONPS-21, at https://www.doi.gov/sites/ doi.gov/files/ fy2020-nps-justification.pdf.

- National Recreation Trails (NRTs) are on federal, state, or private lands that are in, or reasonably accessible to, urban areas. They provide for a variety of outdoor recreation uses.
- Connecting or Side Trails provide public access to the other types of nationally designated trails or connections between such trails.

Congress plays an ongoing role in shaping the National Trails System through legislation and oversight. Congress establishes new trails within the system; directs the Administration to study potential new trails; determines the level of agency funding for trail management; and considers whether new trail categories (such as "national discovery trails") should be included in the system, among other roles. For individual trails, Congress has made specific provisions concerning land acquisition, trail use, and other matters. Ongoing issues for Congress include whether to designate additional trails, how to balance trail designation with other potential land uses, whether trail designation should be accompanied by federal land acquisition, what activities should be permitted on trails, and how to appropriately balance federal and nonfederal funding for trails, among other issues.

BACKGROUND

During the early history of the United States, trails served as routes for commerce and migration. Since at least the early 20th century, trails also have been constructed to provide access to scenic areas. The first interstate recreational trail, now known as the Appalachian National Scenic Trail, was developed in the 1920s and 1930s.[5] In 1945, legislation to establish a

---

[5] An October 1921 journal article proposed the idea of a series of camps along the Appalachian Mountains from New Hampshire to North Carolina, with a trail connecting them. The trail was in place as a continuous footpath by the late 1930s. See Appalachian Trail Conservancy, "About the Trail: History," at http://www.appalachiantrail.org/about-the-trail/history.

"national system of foot trails" was introduced but not enacted.[6] In the years following the Second World War, the nation sought increased opportunities to enjoy the outdoors.[7] In 1965, in a message to Congress on "Natural Beauty," President Lyndon Johnson called for the nation "to copy the great Appalachian Trail in all parts of our country, and make full use of rights-of-way and other public paths."[8] Three years later, the National Trails System Act was enacted.

The system began in 1968 with two scenic trails: the Appalachian National Scenic Trail, which stretches roughly 2,190 miles from Mount Katahdin, ME, to Springer Mountain, GA;[9] and the Pacific Crest National Scenic Trail, which covers roughly 2,650 miles along the mountains of Washington, Oregon, and California.[10] The system was expanded a decade later when Congress designated four historic trails, with more than 9,000 miles, and another scenic trail, along the Continental Divide, with 3,100 miles.[11] Currently, there are a combined 30 NHTs and NSTs covering just under 60,600 miles. (See Table 1 and Figure 1.)[12]

---

[6] The bill, H.R. 2142, 79th Congress, would have amended the Federal-Aid Highway Act of 1944 to provide for a system of foot trails to complement the nation's highway system. For discussion, see Donald D. Jackson, "The Long Way 'Round," *Wilderness*, vol. 51, no. 181 (summer 1998), pp. 19-20; and Sarah Mittlefehlt, "The Tangled Roots of the Appalachian Trail: A Social and Environmental History" (Ph.D. diss, University of Wisconsin, 2008), pp. 150-151.

[7] For example, in 1958, Congress established the Outdoor Recreation Resources Review Commission to make a nationwide study of outdoor national recreation needs. Outdoor Recreation Resources Review Commission, *Outdoor Recreation for America* (Washington, DC: January 1962). This chapter indicated that 90% of all Americans participated in some form of outdoor recreation and that walking for pleasure ranked second among all recreation activities.

[8] *Congressional Record*, vol. 111 (February 8, 1965), p. 2087.

[9] See Appalachian Trail Conservancy, "About the Trail," at http://www.appalachiantrail.org/about-the-trail.

[10] See Pacific Crest Trail Association, "PCT FAQ," at http://www.pcta.org/discover-the-trail/faq/.

[11] P.L. 95-625, National Parks and Recreation Act of 1978.

[12] Personal communication from Peter Bonsall, GIS Specialist, NPS, April 11, 2019. According to NPS, these data are based on GIS analysis of centerline data for all 30 NSTs and NHTs and are not reflective of legislative language designating the trails. *Figure 1* also reflects the recent trail extensions enacted as part of P.L. 116-9. Estimates may differ from publicly available data for these reasons.

Table 1. National scenic and national historic trails, by date of designation

	States	Administering Agency	Date of Designation	Public Law
National Scenic Trails (NSTs)				
Appalachian NST	CT, GA, MA, MD, ME, NC, NH, NJ, NY, PA, TN, VA, VT, WV	NPS	Oct. 2, 1968	P.L. 90-543
Pacific Crest NST	CA, OR, WA	FS	Oct. 2, 1968	P.L. 90-543
Continental Divide NST	CO, ID, MT, NM, WY	FS	Nov. 10, 1978	P.L. 95-625
North Country NST	MI, MN, ND, NY, OH, PA, VT, WI[a]	NPS	Mar. 5, 1980	P.L. 96-199
Ice Age NST	WI	NPS	Oct. 3, 1980	P.L. 96-370
Potomac Heritage NST	DC, MD, PA, VA	NPS	Mar. 28, 1983	P.L. 98-11
Natchez Trace NST	AL, MS, TN	NPS	Mar. 28, 1983	P.L. 98-11
Florida NST	FL	FS	Mar. 28, 1983	P.L. 98-11
Arizona NST	AZ	FS	Mar. 30, 2009	P.L.111-11
New England NST	CT, MA	NPS	Mar. 30, 2009	P.L.111-11
Pacific Northwest NST	ID, MT, WA	FS	Mar. 30, 2009	P.L.111-11
National Historic Trails (NHTs)				
Oregon NHT	ID, KS, MO, NE, OR, WA, WY	NPS	Nov. 10, 1978	P.L. 95-625
Mormon Pioneer NHT	IA, IL, NE, UT, WY	NPS	Nov. 10, 1978	P.L. 95-625
Lewis and Clark NHT	IA, ID, IL, IN, KS, KY, MO, MT, ND, NE, OH, OR, PA, SD, WA, WVb	NPS	Nov. 10, 1978	P.L. 95-625
Iditarod NHT	AK	BLM	Nov. 10, 1978	P.L. 95-625
Overmountain Victory NHT	NC, SC, TN, VA	NPS	Sept. 8, 1980	P.L. 96-344
Nez Perce NHT	ID, MT, OR, WA	FS	Oct. 6, 1986	P.L. 99-445
Santa Fe NHT	CO, KS, MO, NM, OK	NPS	May 8, 1987	P.L. 100-35
Trail of Tears NHT	AL, AR, GA, IL, KY, MO, NC, OK, TN	NPS	Dec. 16, 1987	P.L. 100-192

	States	**Administering Agency**	**Date of Designation**	**Public Law**
Juan Bautista de Anza NHT	AZ, CA	NPS	Aug. 15, 1990	P.L. 101-365
California NHT	CA, CO, ID, KS, MO, NE, NV, OR, UT, WY	NPS	Aug. 3, 1992	P.L. 102-328
Pony Express NHT	CA, CO, KS, MO, NE, NV, UT, WY	NPS	Aug. 3, 1992	P.L. 102-328
Selma to Montgomery NHT	AL	NPS	Nov. 12, 1996	P.L. 104-333
El Camino Real de Tierra Adentro	NM, T)(	NPS & BLM	Oct. 13, 2000	P.L. 106-307
NHT				
Ala Kahakai NHT	HI	NPS	Nov. 13, 2000	P.L. 106-509
Old Spanish NHT	AZ, CA, CO, NM, NV, UT	NPS & BLM	Dec. 4, 2002	P.L. 107-325
El Camino Real de los Tejas NHT	LA, T)(	NPS	Oct. 18, 2004	P.L. 108-342
Captain John Smith Chesapeake	DC, DE, MD, PA, NY, VA	NPS	Dec. 19, 2006	P.L. 109-418
NHT				
Star-Spangled Banner NHT	DC, MD, VA	NPS	May 8, 2008	P.L. 110-229
Washington-Rochambeau Revolutionary Route NHT	CT, DC, DE, MA, MD, NJ, NY, PA, RI, VA	NPS	Mar. 30, 2009	P.L. 111-11

Sources: 16 U.S.C. §§1241-1251; National Park Service, The National Parks: Index, 2012-2016; National Park Service, "National Trails System: Frequently Asked Questions," at https://www.nps.gov/subjects/ nationaltrailssystem/faqs.htm.

Note: NPS = National Park Service; FS = U.S. Forest Service; BLM = Bureau of Land Management.

[a] Reflects a roughly 40-mile extension enacted by P.L. 116-9.

[b] Reflects addition of 1,200 miles as part of the Eastern Legacy Extension enacted by P.L. 116-9.

Additionally, the system contains just under 1,300 NRTs and seven connecting or side trails, including trails in every state, the District of Columbia, and Puerto Rico. The National Trails System Act also authorizes the preservation of abandoned railroad rights-of-way for rails-to-trails conversions (16 U.S.C. §1247).[13]

---

[13] The provisions concerning railroad rights-of-way were added to the act in 1983 by P.L. 98-11.

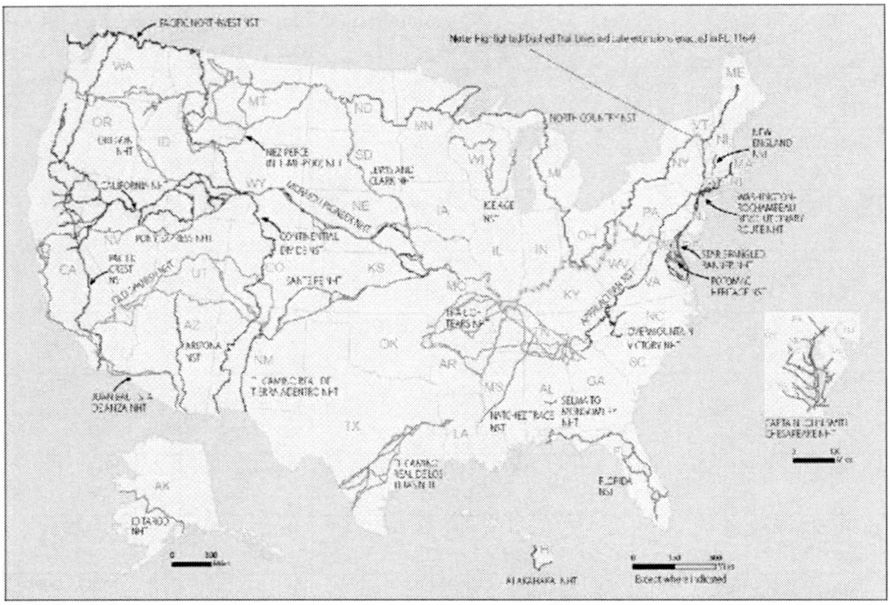

Source: CRS with centerline data adapted from National Park Service, "National Trails System Interactive Map," at https://nps.maps.arcgis.com/home/index.html. Data for trail extensions enacted in P.L. 116-9 are based on cartographic shape files and do not reflect geospatial data. NPS has yet to establish final centerline data for these routes.

Notes: NHT = National Historic Trails; NST = National Scenic Trails. Trail colors are for purposes of distinguishing the trails and do not indicate categories.

Figure 1. Map of National Scenic and National Historic Trails.

DESIGNATION AND LAND ACQUISITION

NSTs and NHTs are designated by acts of Congress.[14] Prior to establishing a trail, Congress typically directs the Secretary of the Interior or the Secretary of Agriculture to study the route for potential inclusion in the system.[15]

---

[14] 16 U.S.C. §1244(a).

[15] 16 U.S.C. §1244(b). The law does not explicitly require that a trail be studied before it is added to the system, but in practice, Congress has directed a prior study for almost all of the national scenic and historic trails. The Secretaries may not undertake trail studies without congressional authorization.

The studies address both the suitability (i.e., characteristics that make the proposed trail "worthy of designation as a national scenic or national historic trail")[16] and the feasibility (i.e., physical and financial viability) of adding the trail to the system. The act contains additional criteria for NHTs, which must (1) be established by historic use and be significant because of that use; (2) be significant with respect to a broad facet of American history, such as trade and commerce, exploration, migration and settlement, or military campaigns; and (3) have significant potential for public recreational use or historical interest.[17]

In contrast to national scenic and historic trails, national recreation trails may be designated by the Secretaries of the Interior and Agriculture with the consent of the federal agency, state, or political subdivision with jurisdiction over the lands involved.[18] Recreation trails must be reasonably accessible to urban areas and must meet other criteria as prescribed by the act or by the Secretaries.[19] The Secretaries also have authority to designate connecting and side trails.[20]

For all four trail types, routes may intersect both federal and nonfederal lands. The law provides limited authorities for federal land acquisition in connection with the trails. Along the designated rights-of-way for NSTs and NHTs, the Secretaries may acquire land in areas that are already under their administrative jurisdiction (e.g., on trail segments that lie within the boundaries of an existing national park or national forest but are not federally owned). Outside their administrative boundaries, the Secretaries are to encourage state and local governments either to acquire trail lands or to enter into agreements with private landowners for the necessary rights-of-way. Only if state and local governments fail to do so may the federal government acquire the land or form cooperative

---

[16] 16 U.S.C. §1244(b)(3).

[17] 16 U.S.C. §1244(b)(11).

[18] Although most of these trails are designated administratively, Congress has also occasionally established national recreation trails. For example, P.L. 110-229 established the Jim Weaver Loop Trail in Oregon's Willamette National Forest as a national recreation trail.

[19] 16 U.S.C. §1243.

[20] 16 U.S.C. §1245.

agreements with landowners.[21] For NRTs, the provisions are more limited, in that federal land acquisition may take place only within existing administrative boundaries.[22] Connecting and side trails may include nonfederal lands only if no federal acquisition is involved.[23]

When adding individual trails to the system, Congress has often included specific land acquisition provisions—for example, authorizing federal acquisition only from willing sellers or establishing a geographical boundary for land acquisition, such as within a quarter-mile on either side of the trail.[24] P.L. 111-11 gave federal land management agencies the authority to purchase land from willing sellers for a number of trails for which Congress had previously prohibited federal land acquisition.

ORGANIZATION AND MANAGEMENT

The 30 national scenic and historic trails are administered by either the Secretary of the Interior or the Secretary of Agriculture, acting through the land management agencies. The NPS administers 21 of the 30 trails; the FS administers 6 trails; the Bureau of Land Management (BLM) administers 1 trail; and the NPS and BLM jointly administer 2 trails.[25] The administering agency typically develops the trail management plan,[26] oversees development of trail segments, coordinates trail marking and mapping, develops maintenance standards, coordinates trail interpretation,

---

[21] 16 U.S.C. §1246(d) and (e). The law authorizes that, when federal acquisition is necessary, it may take place through donations, by purchase with donated or appropriated funds, by exchange, and, within limits, by condemnation (16 U.S.C. §1246(f) and (g)).

[22] 16 U.S.C. §1246(d).

[23] 16 U.S.C. §1245.

[24] For examples of these types of provisions for individual trails, see 16 U.S.C. 1244(a).

[25] See NPS, "Frequently Asked Questions," https://www.nps.gov/subjects/nationaltrailssystem/faqs.htm.

[26] 16 U.S.C. §1244(e) and (f) require the Secretaries to develop management plans for all national scenic and historic trails, in consultation with affected federal agencies, states, and other stakeholders.

administers cooperative and interagency agreements, and provides financial assistance to others for trail purposes, among other functions.[27]

The agencies point to a distinction between trail administration and trail management: While there is usually only one administering agency, multiple federal agencies, state and local governments, private groups, and individuals may own and manage lands along a national scenic or historic trail.[28] The National Trails System Act authorizes the administering Secretary to enter into cooperative agreements with state or local governments, landowners, private organizations, or individuals for trail development, operation, and maintenance.[29] In addition, several federal agencies involved with the trails signed a memorandum of understanding (MOU) in 2006 to coordinate federal trail management.[30] In 2018, the U.S. Court of Appeals for the Fourth Circuit ruled that the laws applicable to the overall trail administrator, rather than the agency or entity managing the underlying land components, would govern the permitting of easements or rights-of-way to cross the Appalachian National Scenic Trail.[31]

---

[27] The memorandum of understanding (MOU) cited in footnote 30, among other sources, describes agency functions generally. Agency-specific guidance includes the following: for the National Park Service, Director's Order 45, *National Trails System*, https://www.nps.gov/policy/DOrders/DO_45.pdf; for the Bureau of Land Management, BLM Manual 6250, *National Scenic and Historic Trail Administration*, BLM Manual 6280, *Management of National Scenic and Historic Trails and Trails Under Study or Recommended as Suitable for Congressional Designation*, and BLM Manual 8353, *Trail Management Areas, Secretarially Designated National Recreation, Water, and Connecting and Side Trails*, all available at https://www.blm.gov/policy/manuals; and for the Forest Service, FS Manual 2300, *Recreation, Wilderness, and Related Resource Management*, Chapter 2350, "Trail, River, and Similar Recreation Opportunities," at http://www.fs.fed.us/cgi-bin/Directives/get_dirs/fsm?2300.

[28] See, e.g., National Park Service, "National Trails System: Frequently Asked Questions," at http://www.nps.gov/nts/nts_faq.html.

[29] 16 U.S.C. §1246(h)(1).

[30] Memorandum of Understanding, "The National Trails System 2017-2027," at https://home.nps.gov/subjects/nationaltrailssystem/upload/National_Trails_System_MOU_2017-2027.pdf. The agencies included NPS, FS, and BLM, as well as the Fish and Wildlife Service and the U.S. Army Corps of Engineers, both of which manage segments of some trails, although they do not serve as primary trail administrators. Additionally, the Federal Highway Administration, which provides transportation funding for trails, participated in the MOU. The MOU encourages cooperation among both federal and nonfederal land managers in activities such as resource inventory and mapping, development of new trail segments and sites, mitigation of resource damage, interpretation, and maintenance.

[31] See *Cowpasture River Pres. Ass'n v. Forest Serv.*, 911 F.3d 150, 180-81 (4th Cir. 2018), *pet. for panel and en banc rehearing denied*, 2019 U.S. App. LEXIS 5532 (4th Cir. Feb. 25, 2019). The three-judge panel ruled that the FS lacked statutory authority to grant a pipeline

In contrast to the NSTs and NHTs, NRTs are typically administered by states, localities, and private organizations, with federal agencies participating when the trails cross federal lands. The National Park Service (NPS) is responsible for the overall coordination of the national recreation trails, including nonfederal trails.[32] Nonfederal trail managers have access to federal training and technical assistance, and are eligible for some types of federal funding.[33]

Connecting or side trails are administered by the Secretary under whose jurisdiction the trail lands fall.[34] The seven existing trails are all administered by the Secretary of the Interior.[35]

The Federal Interagency Council on the National Trails System has discussed a number of management issues facing the agencies that administer the National Trails System.[36]

---

right of way across the Appalachian NST, despite being the managing agency for the segment in question. Per the court's decision, because the overall administration of the Appalachian NST was delegated to NPS as a unit of the National Park System, only NPS has the authority to grant such a right of way, which, subject to the applicable laws of the National Park System, is not permitted.

[32] However, the FS administers national recreation trails within the national forests.

[33] For example, designated national recreation trails may receive funding through the Federal Highway Administration's Recreational Trails Program, administered by the states (see http://www.fhwa.dot.gov/environment/recreational_trails/index.cfm), and are eligible to be considered for support through the Challenge Cost-Share Programs of the NPS, BLM, and FS (see, e.g., https://www.nps.gov/orgs/1837/index.htm).

[34] 16 U.S.C. §1245.

[35] Two connecting or side trails were designated by the Secretary of the Interior in 1990: the 18-mile Timm's Hill Trail in Wisconsin, which connects to the Ice Age National Scenic Trail; and the 186-mile Anvik Connector in Alaska, which connects to the Iditarod National Historic Trail. Another four were designated by the Secretary of the Interior in 2012, all water trails that connect to the water-based Captain John Smith National Historic Trail. For more information, see U.S. Department of the Interior, "Four Rivers in Five States to Make Up Connecting Water Trails," press release, May 16, 2012, at http://www.doi.gov/news/pressreleases/AMERICAS-GREAT-OUTDOORS-Secretary-Salazar-Designates-Captain-John-Smith-Chesapeake-National-Historic-Trail.cfm. The Secretary of the Interior designated a seventh connecting trail in 2015 connecting to the Selma to Montgomery National Historic Trail.

[36] Federal Interagency Council on Trails, *National Trails System: Annual Report for FY2013*, February 2014, p. 3, at http://www.nps.gov/nts/2013%2006MOU%20RPT%207%20Version.pdf. The Federal Interagency Council on the National Trails System (also known as the National Trails System Council) was previously known as the Federal Interagency Council on Trails. The 2017 National Trails System MOU renamed the council and reaffirmed its role as an interagency task force consisting of representatives of the Departments of the Interior and Agriculture with the goal of providing for interagency coordination of administration and management of the NTS.

These issues include tight federal agency budgets, financial constraints among partner groups, inconsistent mapping, aging volunteers, and expansion of energy projects and the transmission grid in ways that affect the trails. Other challenges include the lack of awareness among many Americans of the system and its health, community, economic, and educational benefits.

TRAIL USES

The administering Secretary may regulate the use of federally owned portions of the national trails, in consultation with relevant agencies.[37] The Secretary may permit uses "which will not substantially interfere with the nature and purposes of the trail."[38] Such uses may include but are not limited to bicycling, cross-country skiing, day hiking, equestrian activities, jogging or similar fitness activities, overnight and long-distance backpacking, snowmobiling, and surface water and underwater activities.[39] The use of motorized vehicles by the general public is typically prohibited on national scenic trails.[40] However, motorized vehicles may be allowed on national historic trails if they do not substantially interfere with the nature and purposes of the trail and were allowed by administrative regulations at the time of designation.[41]

Trail uses on nonfederal lands—whether segments of scenic, historic, recreation, or connecting trails—are typically controlled at the state and

---

[37] 16 U.S.C. §1246 (i).

[38] 16 U.S.C. §1246 (c).

[39] 16 U.S.C. §1246 (j)

[40] Ibid. However, this provision directs the Secretary to allow motorized vehicle use in certain circumstances, such as for emergencies and when necessary to give adjacent landowners reasonable access to their lands or timber rights. Additionally, specific provisions for the Continental Divide National Scenic Trail (16 U.S.C. §1244(a)(5)) allow motorized use in accordance with regulations established by the administering Secretary.

[41] 16 U.S.C. §1246 (c).

local levels. State, local, and private-sector trail managers may work together to develop cooperative principles for use and management.[42]

FUNDING

Each agency with management authority over national trails has its own funding for carrying out activities related to trail administration and management.[43] Since 2006, federal land management agencies have agreed, within the limits of agency authorities, to eliminate duplicate efforts and increase effectiveness by coordinating requests for and obligation of funds for the National Trails System.[44] Since 1992, the Department of Transportation, through federal transportation programs authorized by Congress,[45] has provided just under $14 billion for bicycle and pedestrian transportation projects, including many transportation trails.[46] Additional sources of funding for trails have included challenge cost-share projects, cooperative agreements with trail partner organizations, charitable foundations, corporations, permits and fees, local excise taxes, and dedicated funds.

---

[42] See, e.g., NPS, Appalachian Trail Project Office, *Comprehensive Plan for the Protection, Management, Development and Use of the Appalachian National Scenic Trail*, 1987, at http://www.nps.gov/appa/parkmgmt/upload/CompPlan_web.pdf.

[43] A total amount of federal appropriations for the National Trails System is not available, since many of the agencies do not separately track trail funding.

[44] Memorandum of Understanding 06-SU-11132424-196, "The National Trails System," at http://www.nps.gov/nts/memorandum2006.html. This agreement was reiterated in the 2017 updated interagency MOU "The National Trails System 2017-2027," at https://home.nps.gov/subjects/nationaltrailssystem/upload/National_Trails_System_MOU_2017-2027.pdf.

[45] Congress has authorized recreational trail assistance in surface transportation laws; for more information, see CRS Report R44388, *Surface Transportation Funding and Programs Under the Fixing America's Surface Transportation Act (FAST Act; P.L. 114-94)*, coordinated by Robert S. Kirk.

[46] See Federal Highway Administration (FHWA), "Federal-Aid Highway Program for Pedestrian and Bicycle Facilities and Programs, FY1992 to FY2018 Obligations," at http://www.fhwa.dot.gov/environment/bicycle_pedestrian/funding/bipedfund.cfm. The agency does not break out the portion of this funding that has gone to trails within the National Trails System. For example, funding in FHWA's Recreational Trails Program (a subset of the agency's overall funding for pedestrian and bicycle transportation) is used for recreational trails both within and outside the system.

LEGISLATION

In the 116th Congress, the omnibus public lands bill, P.L. 116-9, made a number of amendments to the National Trails System. Among the changes, the bill adjusted the boundaries of the North Country National Scenic Trail, extended the Lewis and Clark National Historic Trail, and directed the study of a proposed Pike National Historic Trail running from Missouri through Louisiana. Other bills have been introduced to either designate additional trails or make broader adjustments to the trails system. H.R. 434 would direct the study and designation of the Emancipation National Historic Trail, a route from Galveston, TX, to Freedmen's Town and Emancipation Park in Houston following the migration route taken by newly freed slaves in the 19th century. Another proposal (H.R. 726) would add a new type of trail—"national discovery trails"—to the system. National discovery trails would be extended, continuous interstate trails that provide for outdoor recreation and travel and that connect representative examples of America's trails and communities.[47] These and other 116th Congress bills affecting the National Trails System are shown in Table 2. Many comparable bills had also been introduced in previous Congresses.

In addition to bills introduced in the 116th Congress, other recent developments have included the completion of trail studies directed in earlier legislation. For example, P.L. 111-11 directed the Secretary of the Interior to conduct feasibility studies for both the Chisholm and Great Western trails for potential designation as national historic trails. NPS subsequently completed a feasibility study in 2016 with the recommendation that the trails be designated as the Chisholm National Historic Trail and Great Western National Historic Trail and administered

[47] H.R. 726 would establish the American Discovery Trail, extending from Delaware to California, as the first national discovery trail. Similar bills to create national discovery trails within the system were introduced in previous Congresses. H.R. 3251 in the 115th Congress would also have established this trail type and designated the American Discovery Trail.

as a combined unit of the National Trails System.[48] To date, these recommendations have not resulted in legislation.

Table 2. National Trails System Bills introduced in the 116th Congress

Title or Trail	Type	Bill No.	Status
John D. Dingell, Jr. Conservation, Management, and Recreation Act[a]	Extend/Study	S. 47	Became P.L. 116-9
		H.R. 1727	Introduced
Complete America's Great Trails[b]	Other	S. 809	Introduced
Emancipation National Historic Trail (TX)[c]	Designate	H.R. 434	Hearing Held
National Discovery Trails[d]	Other	H.R. 726	Introduced
		H.R. 2090	Introduced
National Scenic Trails Parity[e]	Other	S. 1027	Introduced
		H.R. 1216	Enacted in P.L. 116-9
North Country National Scenic Trail[f]	Extend	S. 96	Enacted in P.L. 116-9
Lewis and Clark National Historic Trail[g]	Extend	S. 167	Enacted in P.L. 116-9
Pike National Historic Trail[h]	Study	S. 57	Enacted in P.L. 116-9

Source: Table compiled by the Congressional Research Service.

[a]P.L. 116-9, among other purposes, extended both the North Country NST and the Lewis and Clark NHT and directed the study of a proposed Pike NHT running from Missouri through Louisiana.

[b]H.R. 1727 and S. 809 would amend the Internal Revenue Code to allow a tax credit for the fair market value of any NST conservation contribution. The bills would also require the Secretary of the Interior to study the efficacy of such a tax credit in completing, extending, and increasing the number of NSTs.

[c]H .R. 434 would direct the study and designation of the Emancipation National Historic Trail, a route from Galveston, TX, to Freedmen's Town and Emancipation Park in Houston following the migration route taken by newly freed slaves in the 19th century.

[d]H.R. 726 would establish a new trail type, national discovery trails. These extended, continuous interstate trails would be located so as to provide for outdoor recreation and travel, and would connect representative examples of America's trails and communities.

[e]H.R. 2090 and S. 1027 would require that the North Country, Ice Age, and New England NSTs be administered as units of the National Park System.

[f]H.R. 1216 and S. 96 would extend the North Country NST into VT.

[g]S. 167 would extend the Lewis and Clark NHT from Wood River, IL, to Pittsburgh, PA, passing through portions of IA, KY, OH, PA, and WV.

[h]S. 57 would direct NPS to study the feasibility of designating 3,664 miles of routes traveled by Lieutenant Zebulon Pike—extending from MO through portions of KS, NE, CO, NM, and TX, and ending in LA—as a national historic trail.

---

[48] NPS, "Chisholm and Great Western National Historic Trail Feasibility Study/Environmental Assessment" at https://parkplanning.nps.gov/projectHome.cfm?projectID=30803.

In: Key Congressional Reports for June 2019 ISBN: 978-1-53616-662-0
Editor: Stephanie A. Betts

Chapter 7

MAINTAINING ELECTRIC RELIABILITY WITH WIND AND SOLAR SOURCES: BACKGROUND AND ISSUES FOR CONGRESS*

Ashley J. Lawson

SUMMARY

The share of wind and solar power in the U.S. electricity mix grew from 1% in 2008 to 8% in 2018. Wind and solar are variable renewable energy (VRE) sources. Unlike conventional sources, weather variability creates uncertainty about the availability of VRE sources. This uncertainty could potentially result in a lack of reliability.

Some Members of Congress have expressed concerns about the reliability of the electric power system given recent growth in generation from wind and solar sources and projections that growth will continue. According to official metrics, electric reliability was generally stable or improving over the 2013-2017 period. In other words, generation from

* This is an edited, reformatted and augmented version of Congressional Research Service Publication No. R45764, dated June 10, 2019.

wind and solar sources does not appear to be causing electric reliability issues at the national level over this period. Questions remain, however, about maintaining reliability if generation from wind and solar should increase above current projections, as some Members of Congress have supported. Entities in the electric power sector and their regulators are evaluating changes to their approaches to reliability to prepare for this possibility. Congress might seek clarification on whether new or modified approaches are required.

Under the current regulatory framework, the federal government oversees reliability for the generation and transmission systems of the electric power sector. These components comprise the bulk power system and include large-scale wind and solar sources. The Energy Policy Act of 2005 (EPACT05; P.L. 109-58) authorized the Federal Energy Regulatory Commission (FERC) and the North American Electric Reliability Corporation (NERC) to develop and enforce mandatory reliability standards for the bulk power system. Small-scale wind and solar sources, such as rooftop solar photovoltaic (PV) panels, are connected to the distribution system which is localized and under state jurisdiction. Federal mandatory reliability standards do not apply to the distribution system.

The colloquial definition of reliability is "having power when it is needed," but regulators and operators of power system components require a more precise statement of objectives and metrics. FERC and NERC have developed numerous technical standards to address reliability. These standards apply over the range of timescales over which reliability is measured, from milliseconds to years. FERC has approved approximately 100 reliability standards to date, and new standards are developed as needed to respond to changing conditions, including increasing generation from wind and solar sources. Multiple entities spanning multiple jurisdictions work together to maintain electric reliability.

For economic reasons, wind and solar sources tend to be utilized to the maximum extent possible. When their availability changes, which can happen quickly, other sources must quickly respond to maintain reliability. Typically, other sources respond by increasing or decreasing their output, an operation known as balancing. Multiple types of electricity sources are used to balance wind and solar, including some fossil fuel-fired generators, some nuclear generators, other wind and solar sources (provided sufficient transmission availability), energy storage, and demand response. Each of these has benefits and limitations. Some sources and system operations that currently support balancing have received federal financial support in the past, such as tax credits, grants to states or other entities, and Department of Energy research programs. Congress might consider continuing or expanding such support, if lawmakers believed current activities affecting reliability were insufficient.

Beyond developing and enforcing reliability standards, other federal government activities affect electric reliability. For example, FERC's regulation of interstate electricity transmission can be a key determinant of how effectively different electricity sources can meet demand. FERC's regulation of the wholesale electricity markets that operate in some regions of the country may also affect reliability, because market rules can influence which individual generators are used for system balancing. Market prices directly affect project revenues, influencing the kinds of sources that are developed. Additionally, some projects and programs Congress funds support reliability by enabling technology development and providing financial support for projects that support reliability.

INTRODUCTION

Economic factors, new technologies, environmental concerns and associated regulatory policies, and other developments are changing the energy sources used to generate electricity in the United States. One notable change is increased generation from variable renewable energy (VRE) sources such as wind and solar. According to the U.S. Energy Information Administration (EIA), combined generation from wind and solar sources increased from 1% of total electricity generation in 2008 to 9% of total electricity generation in 2018.[1] These sources have weather-dependent availability, meaning that changing weather patterns can change available electricity supply from those sources. In contrast, conventional sources for electricity generation, such as coal, natural gas, or nuclear energy, are usually available under normal weather conditions. Power system operators have adjusted existing reliability standards and planning practices to accommodate weather-dependent wind and solar sources. Further adjustments are being discussed as generation from wind and solar sources continue to grow.

---

[1] CRS analysis of data from EIA, "Electricity Data Browser," accessed April 16, 2019, https://www.eia.gov/electricity/data/browser/. EIA began collecting data on generation from small-scale solar photovoltaic sources in 2014. The statistic for 2018 reported here includes both large-scale and small-scale solar sources.

Congress required the setting and enforcement of electric reliability standards in the Energy Policy Act of 2005 (EPACT05; P.L. 109-58).[2] These standards are developed by the North American Electric Reliability Corporation (NERC) and approved by the Federal Energy Regulatory Commission (FERC) in the United States.[3] These mandatory standards apply to the bulk power system, which is comprised mostly of large-scale generators and electricity transmission systems. Small-scale generators (e.g., rooftop solar electricity generation), publicly owned utilities, and local electricity distribution systems are generally under the jurisdiction of state public utility regulatory commissions (PUCs).

To date, generation from wind and solar sources does not appear to be causing electric reliability issues at the national level. NERC's 2018 annual report on reliability showed that, of the 13 metrics it uses to assess reliability, 9 were stable or improving over the 2013-2017 period and 4 showed trends that were, at least partly, inconclusive. Of the four metrics with inconclusive trends, three improved over this period for a subset of bulk power system components.[4] Data from NERC also indicate that reliability performance is currently stable in regions such as the Midwest and California where the shares of generation from wind and solar sources are above the national average. Questions remain about how higher levels of generation from wind and solar sources might impact electric reliability moving forward.

This chapter provides background on reliability planning in the United States with an emphasis on the effects of daily and seasonal variability in wind and solar sources on the bulk power system. Members of Congress

---

[2] 16 U.S.C. §824o.

[3] Due to the integrated nature of the bulk power system, NERC oversees electric reliability for parts of Canada and Mexico as well. The electric industry originally established NERC's predecessor, the National Electric Reliability Council, in 1968 to coordinate voluntarily reliability efforts. Alaska and Hawaii are exempt from NERC's jurisdiction.

[4] The planning reserve margin in parts of Texas was an exception to the generally positive performance against reliability metrics. This metric relates to system planning over several years, and does not necessarily indicate reliability disruptions will occur. The system operator for most parts of Texas has operational tools in place to account for low planning reserve margins. NERC, *State of Reliability 2018*, June 2018, Table 3.1, https://www.nerc.com/pa/RAPA/PA/Performance%20Analysis%20DL/NERC_2018_SOR_06202018_Final.pdf.

might consider how reliability could be impacted if generation from wind and solar sources increases, as many analysts expect. Other reliability concerns, such as cyber and physical security, small-scale generators, and local distribution networks, may be of interest to Congress but are not discussed at length in this chapter.

ELECTRIC POWER SECTOR OVERVIEW

As shown in Figure 1 the electric power sector consists primarily of three systems. The generation system consists of power plants that generate electricity. The transmission system consists of high voltage transmission lines that move power across long distances. The distribution systems make final delivery of electricity to homes and businesses. This chapter will refer to the combined generation and transmission systems as the bulk power system, following the definition Congress established in EPACT05:

> The term "bulk-power system" means—(a) facilities and control systems necessary for operating an interconnected electric energy transmission network (or any portion thereof); and
>
> (b) electric energy from generation facilities needed to maintain transmission system reliability.
>
> The term does not include facilities used in the local distribution of electric energy.[5]

Notably, the discussion in this chapter generally excludes distributed energy resources such as rooftop solar electricity generation. These resources might pose separate reliability challenges that Congress might choose to consider.

[5] 16 U.S.C. §824o.

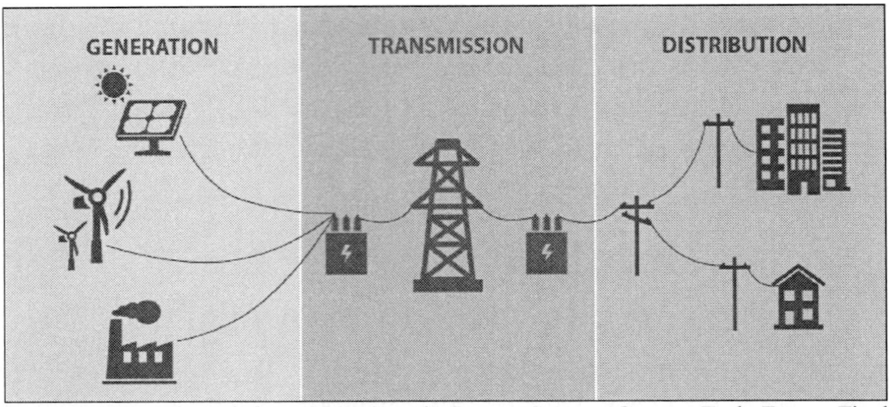

Source: CRS, adapted from U.S.-Canada Power System Outage Task Force, Final Report on the August 14, 2003, Blackout in the United States and Canada: Causes and Recommendations, April 2004, p. 5, https://www.energy.gov/sites/prod/files/oeprod/DocumentsandMedia/BlackoutFinal-Web.pdf.

Note: Not all types of components in each system are shown.

Figure 1. Simplified Schematic of Electric Power Sector Systems.

Ownership structures for bulk power system components vary across the country. In some regions, shown in *Figure 2*, competitive markets exist for wholesale electric power, and regional transmission organizations (RTOs) and independent system operators (ISOs) manage the generation and transmission components of the power system.[6] In RTO regions, electricity generators compete to sell power to distribution utilities. The RTO manages an auction process to select the sources for generation that distribution utilities resell to end-use customers. The RTO is also responsible for managing the transmission system and overseeing reliability within its boundaries. In RTO regions, market signals primarily determine investment decisions. Some RTOs operate separate auction processes specifically for essential reliability services.[7] According to FERC, two-thirds of U.S. electricity demand comes from RTO regions.[8]

---

[6] A more detailed history of RTOs and ISOs, and further description of the functions summarized in this paragraph are in CRS Report R44783, *The Federal Power Act (FPA) and Electricity Markets*, by Richard J. Campbell, and CRS Report R43093, *Electricity Markets—Recent Issues in Market Structure and Energy Trading*, by Richard J. Campbell.

[7] The term essential reliability services refers to a group of functions that generators perform, usually automatically, over timescales of seconds or less. Damage to bulk power system

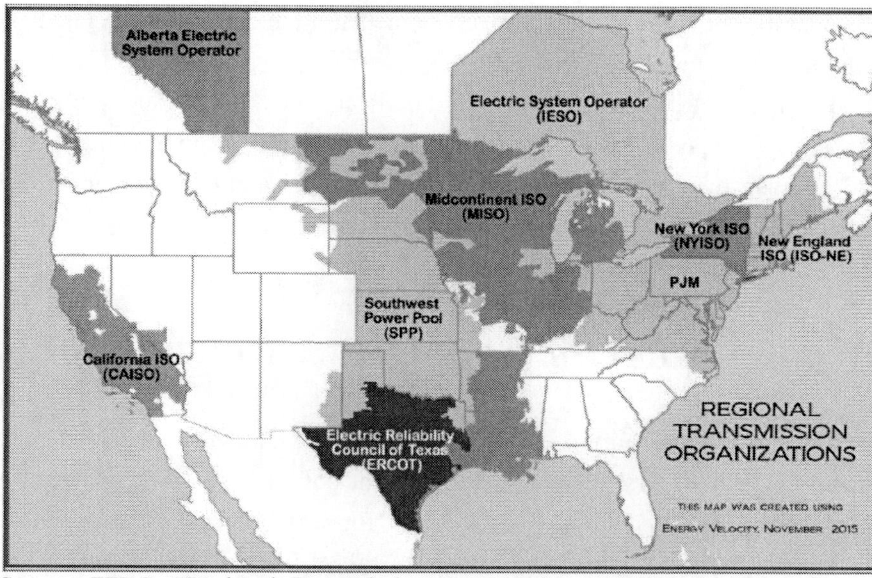

Source: FERC, "Regional Transmission Organizations (RTO)/Independent System Operators (ISO)," October 18, 2018, update, https://www.ferc.gov/industries/electric/indus-act/rto.asp.

Notes: RTOs manage almost two-thirds of U.S. electricity generation, including day-to-day reliability responsibilities. In non-RTO regions, shown as white in this map, vertically-integrated utilities, municipal utilities, or electric co-operatives have these functions. ISO = Independent System Operator. Alaska and Hawaii are not shown because RTOs and ISOs do not operate in those states.

Figure 2. Map of Regional Transmission Organizations (RTOs).

In non-RTO regions, vertically integrated electric utilities are largely responsible for power generation, transmission, and distribution of electricity to end-use customers. These utilities are regulated as natural monopolies and, unlike utilities in RTO regions, do not face competition for generation and transmission services. These utilities may also take responsibility for some aspects of reliability as discussed in the Appendix. State regulators generally oversee these utility operations and are

---

components could occur if sufficient essential reliability services are not available. More details are provided in the *Appendix*.

[8] FERC, *Electric Power Markets: National Overview*, updated April 13, 2017, https://www.ferc.gov/market-oversight/mkt-electric/overview.asp.

responsible for authorizing new investments, including those related to reliability.[9]

Even in RTO regions, municipal utilities and rural electric cooperatives may own generation and transmission system components and oversee their operation.[10] These systems and operation are generally outside of federal and state regulatory jurisdiction.

WHAT IS ELECTRIC RELIABILITY?

A colloquial definition of electric reliability is "having power when it is needed." Operators of bulk power system components, though, require specific and highly technical definitions for reliability. For purposes of regulation, these definitions are provided in the form of NERC reliability standards. NERC develops individual standards for each set of power system components, which may include separate standards covering different reliability timescales for each set of components. As NERC defines "reliability standard," it

> includes requirements for the operation of existing Bulk-Power System facilities, including cybersecurity protection, and the design of planned additions or modifications to such facilities to the extent necessary to provide for Reliable Operation of the Bulk-Power System, but the term does not include any requirement to enlarge such facilities or to construct new transmission capacity or generation capacity.[11]

---

[9] Federal Power Marketing Administrations are exceptions. For more details see CRS Report R45548, *The Power Marketing Administrations: Background and Current Issues*, by Richard J. Campbell.

[10] For further discussion of municipal utilities and rural electric cooperatives see DOE, *United States Electricity Industry Primer*, July 2015, https://www.energy.gov/sites/prod/files/2015/12/f28/united-states-electricity-industryprimer.pdf.

[11] NERC, *Glossary of Terms Used in NERC Reliability Standards*, July 3, 2018, p. 26, https://www.nerc.com/pa/Stand/Glossary%20of%20Terms/Glossary_of_Terms.pdf. This definition of reliability standard was established in the Energy Policy Act of 2005 (P.L. 109-58) from which NERC derives its authority. See 16 U.S.C. §824o.

Distribution System Reliability

As shown in *Figure 1*, the electric distribution system makes final delivery of electric power to businesses, households, and other consumers. Voltage levels on distribution systems are lower than on transmission systems, but otherwise many distribution system components operate in the same way as transmission systems. Different ownership models for electric distribution utilities include investor-owned, publicly owned, and cooperative. In some cases, the distribution utility also owns generation and transmission assets, but in other cases it does not. Regardless of whether distribution utilities also own and operate generation and transmission system components, they are recognized as natural monopolies and regulated by states.

Reliability of distribution systems has not been traditionally viewed as a federal issue, since the effects of outages in distribution systems are localized. Nonetheless, distribution system outages can be impactful.[12] Past federal efforts to improve distribution system reliability include funding for Smart Grid upgrades and grants to communities to install microgrids or other electricity projects.[13]

Small-scale wind and solar generators are frequently located on the distribution system. Often, these occur in so-called behind-the-meter configurations in which the generator is on the customer side of the distribution utility's meter. These sources generally cannot be controlled by system operators the way sources on the transmission system can be, raising reliability concerns. An emerging proposal to account for this situation is an independent distribution system operator (IDSO) that could balance generation and electricity demand in the distribution system. The IDSO would be analogous to ISOs and RTOs in the transmission system.[14]

When all bulk power system components meet reliability standards, NERC expects the vast majority of individuals to have the full amount of electricity they desire. NERC reliability standards do not apply to local electricity distribution system components and operations (see discussion in text box, "Distribution System Reliability"), so localized outages could still occur when reliability standards are met. An analysis found that from

---

[12] For a discussion of how distribution system outages can be impactful, see CRS Report R42696, *Weather-Related Power Outages and Electric System Resiliency*, by Richard J. Campbell.

[13] Definitions for microgrids vary, but they are generally understood to include sources of electricity supply and demand within defined electrical boundaries. Some are capable of operating in isolation of the bulk power system. Lawrence Berkeley National Laboratory, *Microgrid Definitions*, April 26, 2019, https://building-microgrid.lbl.gov/ microgrid-definitions.

[14] The IDSO concept is described in Farrokh Rahimi and Sasan Mokhtari, "From ISO to DSO," *Public Utilities Fortnightly*, June 2014, and James Tong and Jon Wellinghoff, "Rooftop Parity: Solar for Everyone, Including Utilities," *Public Utilities Fortnightly*, August 2014.

2008 to 2014, upwards of 90% of power outages originated in local distribution systems.[15]

This measure includes major events (e.g., hurricanes), but may not capture the full scope or severity of large-scale outages. NERC's reliability standards are meant to ensure an Adequate Level of Reliability (ALR) for the bulk power system during normal operating conditions and following localized disturbances such as lightning strikes.[16]

For economic reasons, some risk of occasional power loss is accepted in reliability planning. A common goal is to limit outages to no more than 1 day every 10 years under normal operating conditions.[17]

Achieving ALR is not the same goal as preventing all brownouts and blackouts. Bulk power system outages could still occur when reliability standards are fully met. These outages might follow a major event such as a hurricane affecting large areas of the bulk power system.

Generally, factors that increase uncertainty reduce reliability, and factors that reduce uncertainty increase reliability. Wind and solar are types of variable renewable energy sources of electricity, and weather is a key source of uncertainty for forecasts of generation from these sources.

---

[15] Researchers from the Institute of Electric and Electronics Engineers (IEEE) Distribution Reliability Working Group and the U.S. Energy Information Administration (EIA) found different values depending on whether outages are measured in terms of duration or frequency, whether major events are included, the which methodology is used for attributing an outage's origination to either the bulk power system or the distribution system. All measures of customer-weighted average values showed at least 90% of outages originated in the distribution system. The mean and median values, however, sometimes showed a lower share (as low as 83%) of outages originating in the distribution system. Joseph Eto et al., "Distribution System Versus Bulk Power System: Identifying the Source of Electric Service Interruptions in the US," *IET Generation, Transmission & Distribution*, vol. 13, no. 5 (2019). These findings are consistent with other analyses that the vast majority of power outages occur in distribution systems.

[16] While cybersecurity is an important aspect of power system reliability that NERC addresses, it is outside the scope of this chapter. That topic is discussed in CRS Report R43989, *Cybersecurity Issues for the Bulk Power System*, by Richard J. Campbell.

[17] This target is called loss of load expectation (LOLE) and is calculated using estimates of generation, demand, and the probabilities of outages on system components. The 1 outage day every 10 years target is equivalently expressed as 0.1 days/year. A technical discussion of LOLE methods in the context of variable sources such as wind and solar is available in NERC, *Methods to Model and Calculate Capacity Contributions of Variable Generation for Resource Adequacy Planning*, March 2011, https://www.nerc.com/files/ivgtf1-2.pdf.

Electricity System Resilience

An emerging topic of interest is the concept of electricity system resilience. This term is somewhat similar to reliability in that both terms suggest that power will be available when it is wanted. The term resilience, however, tends to be used in a broader sense and often in the context of disasters. As the National Academies noted

> Resilience is not just about lessening the likelihood that these outages will occur. It is also about limiting the scope and impact of outages when they do occur, restoring power rapidly afterwards, and learning from these experiences to better deal with events in the future.[18]

Whereas the electric power industry has developed technical standards and metrics for reliability that are suitable for regulatory purposes, a technical definition of resilience is still under development. FERC initiated Docket No. AD18-7-000 on January 8, 2018,

> (1) to develop a common understanding among the Commission, industry, and others of what resilience of the bulk power system means and requires; (2) to understand how each RTO and ISO assesses resilience in its geographic footprint; and (3) to use this information to evaluate whether additional Commission action regarding resilience is appropriate at this time.[19]

This chapter does not discuss power system resilience in depth, although some system operations and other activities that support reliability might also support power system resilience. This chapter uses the terms reliability and reliability standard as those terms are currently used by NERC.

In contrast, conventional sources such as coal and nuclear have long-lasting, on-site fuel supplies that reduce the uncertainty about their availability.[20] This difference has raised questions about how to integrate large amounts of VRE sources into the existing bulk power system, since it

---

[18] National Academies of Sciences, Engineering, and Medicine, *Enhancing the Resilience of the Nation's Electricity System*, The National Academies Press, 2017, p. 1, https://doi.org/10.17226/24836.

[19] FERC, *Order Terminating Rulemaking Proceeding, Initiating New Proceeding, and Establishing Additional Procedures*, January 8, 2018, Para. 18, https://ferc.gov/CalendarFiles/20180108161614-RM18-1-000.pdf.

[20] On-site coal stockpiles can typically supply 2-3 months of continuous generation. Nuclear fuel rods are typically used for up to six years before being removed. Refueling cycles of 18 or 24 months are common, during which a portion of the reactor's fuel rods are replaced. DOE, *Staff Report to the Secretary on Electricity Markets and Reliability*, 2017, https://www.energy.gov/downloads/download-staff-report-secretary-electricity-markets-andreliability.

was not originally designed to accommodate large amounts of weather-dependent sources of electricity.[21]

Figure 3 shows typical patterns for electricity generation for wind and solar sources in the United States. Wind generation tends to peak overnight and during winter months.[22]

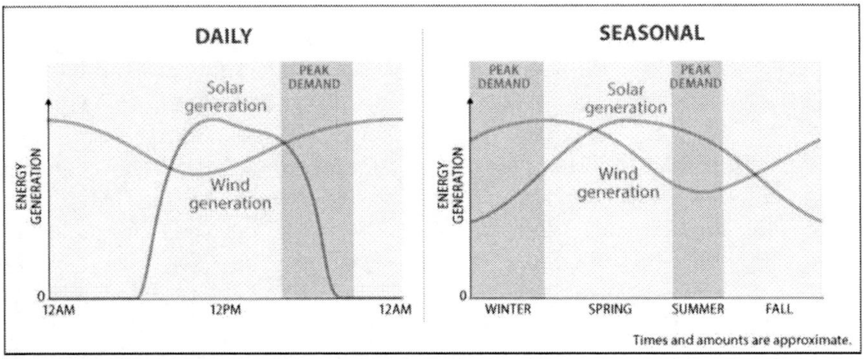

Sources: CRS. Daily generation profiles based on CAISO, Renewables Trend, May 13, 2019, http://www.caiso.com/TodaysOutlook/Pages/supply.aspx. Daily peak demand based on national data from EIA, "U.S. Electric System Operating Data," https://www.eia.gov/realtime_grid/#/summary/demand. Seasonal profiles and peak demand based on 2018 data for total United States, EIA, "Electricity Data Browser," https://www.eia.gov/electricity/data/browser/.

Notes: The y-axis represents the potential energy generation from each individual source, from a minimum of zero to a maximum defined by the installed capacity of the source. Actual generation from any individual source will depend on local weather conditions and other factors. Wind seasonal generation patterns vary by region, so the pattern shown here may not apply in all cases. The daily pattern for solar is typical of south-facing photovoltaic panels in the United States. West-facing panels tend to reach maximum generation 1-2 hours later in the day.

Figure 3. Generation Patterns for U.S. Wind and Solar Sources.

---

[21] Many discussions of the challenges surrounding use of VRE exist. See, for example, Department of Energy, *Staff Report to the Secretary on Electricity Markets and Reliability*, 2017, https://www.energy.gov/downloads/downloadstaff-report-secretary-electricity-markets-and-reliability.

[22] Hawaii and parts of California, Oregon, and Washington tend to have maximum generation from wind sources during summer months due to unique meteorological circumstances. Pacific Northwest Laboratory, *Wind Energy Resource Atlas of the United States*, 1986, https://rredc.nrel.gov/wind/pubs/atlas/chp2.html.

Solar generation, on the other hand, tends to be highest during the middle of the day and during the summer. Though these typical patterns are well established for most of the United States, actual generation from wind and solar sources at any particular moment will depend upon specific weather conditions.

CHANGING ELECTRICITY GENERATION PROFILE

The electric power sector is increasing its use of sources associated with more uncertainty in availability. According to the U.S. Energy Information Administration, combined generation from wind and utility-scale solar sources increased from 1% of total electricity generation in 2008 to 8% of total electricity generation in 2018.[23] Of the generation in 2018 from wind and utility-scale solar sources, 80% came from wind. Conventional sources such as coal, natural gas, and nuclear comprised a large majority of generation over this time period. The annual share of generation from different sources from 2008 to 2018 in shown in *Figure 4*.

National-level data are not indicative of how generation from wind and solar sources varies across the country. Similarly, annual data do not show how electricity generation varies throughout the day or during different seasons. For example, during brief periods in some regions, wind and solar sources have provided a majority of the energy for electricity generation. Some examples are

- Generation from wind sources supplied 56% of electricity demand in ERCOT, the RTO covering most of Texas, at 3:10 am on January 19, 2019.[24]

---

[23] CRS analysis of data from EIA, "Electricity Data Browser," accessed April 16, 2019, https://www.eia.gov/electricity/data/browser/. EIA began collecting data on generation from small-scale solar photovoltaic (PV) sources in 2014. When small-scale solar PV is included in the total, wind and solar made up 9% of total generation in 2018. Typically, small-scale solar PV sources are connected to the distribution system, not the bulk power system.

[24] Jeff Thou, *ERCOT Sets Record Wind Output and Penetration Rate over the Holiday Weekend*, S&P Global Platts, January 22, 2019, https://www.spglobal.com/platts/en/market-

- Generation from solar sources supplied 59% of electricity demand in CAISO, the RTO covering most of California, at 2:45 pm on March 16, 2019.[25]
- Generation from wind supplied 67.3% of electricity demand in SPP, the RTO covering many central states, at 1:25 am on April 27, 2019.[26]

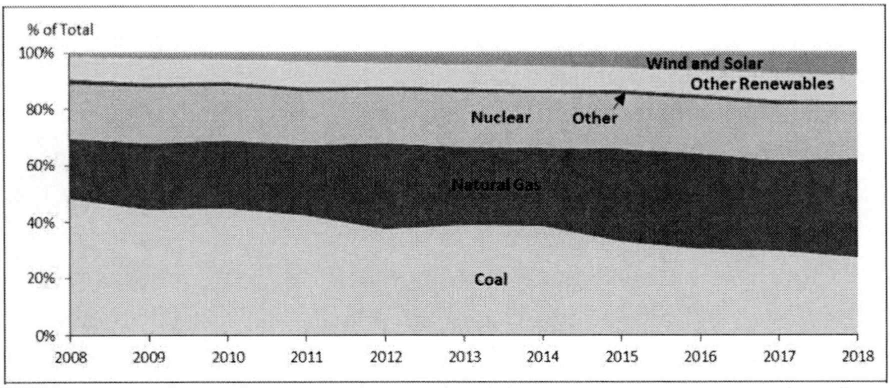

Source: CRS analysis of data from EIA, "Electricity Data Browser," accessed April 16, 2019, https://www.eia.gov/electricity/data/browser/.

Notes: Some EIA categories are combined. Other = petroleum liquids, petroleum coke, other gas, and other; Other Renewables = wood and wood-derived fuels, landfill gas, biogenic municipal solid waste, other waste biomass, geothermal, conventional hydroelectric, and hydroelectric pumped. "Solar" does not include generation from small-scale solar because data were not collected before 2014.

Figure 4. Annual Net U.S. Electricity Generation by Source, 2008-2018. (Percent of total)

These events all set records for maximum share of generation from renewable sources, and the bulk power system maintained reliability during them. Some advocates for increased use of wind and solar sources

insights/latest-news/electric-power/012219-ercot-setsrecord-wind-output-and-penetration-rate-over-the-holiday-weekend.

[25] Christian Roselund, "California Sets a New Solar Output Record—and It Isn't Even Spring Yet," *PV Magazine*, March 18, 2019, https://pv-magazine-usa.com/2019/03/18/california-sets-a-new-solar-output-record-and-it-isnt-evenspring-yet/.

[26] SPP, "SPP set a new #wind-penetration record of 67.3% at 1:25 a.m., April 27," Twitter, April 29, 2019, https://twitter.com/SPPorg/status/1122926771761025024.

have pointed to events like these as evidence that VRE sources can be used to an even greater degree without impacting reliability.[27]

Extrapolating these events to scenarios of correspondingly high national levels of generation from wind and solar sources, however, is complicated by several factors. First, these events were all short lived, typically five minutes or less. Further, these events all occurred when electricity demand was relatively low, namely weekend days during cool months. During times of the year when electricity demand is high, such as the summer cooling season, the share of electricity generation from renewable sources is lower.[28] For example, SPP has reported that during its peak demand hours in 2016, wind supplied 11% of generation while conventional sources such as coal (47%) and natural gas (33%) supplied the majority of electricity.[29] The seasonality of VRE availability also likely contributed to these record-setting events, especially for wind, which tends to have maximum generation during winter and spring months.

BALANCING VARIABLE RENEWABLE ENERGY

Electricity is essentially generated as a just-in-time commodity, due to limited energy storage capacities. If electricity supply and demand differ by too much, system components could be damaged, leading to system

---

[27] For example, a wind industry trade association made this argument after similar records were broken in 2016. Michael Goggin, *The Records Keep Falling: More New Highs in Wind Energy Output*, AWEA, February 23, 2016, https://www.aweablog.org/the-records-keep-falling-more-new-highs-in-wind-energy-output/.

[28] VRE sources are generally used to the maximum extent possible, for economic reasons, as described in the section "Balancing Variable Renewable Energy." Unlike other types of sources, the potential supply of electricity from VRE sources does not usually change in response to total electricity demand. Thus, when demand increases and more electricity in total is generated to meet that demand, the relative share of generation from VRE sources decreases. If total demand for electricity at any moment is 10,000 MW and 5,000 MW of electricity from VRE sources is available, the share of generation from VRE sources will likely be 50%. If, however, total demand were 50,000 MW, the share of generation from VRE sources would likely be 10%.

[29] SPP Supply Adequacy Working Group, *Wind and Solar Report*, May 23, 2017, https://www.spp.org/documents/53721/sawg%20approved_wind%20and%20solar%20report.pdf.

instability or potential failure. The operations that keep electricity supply and demand within acceptable levels are known as balancing.

Balancing involves increasing or decreasing output from generators according to system conditions over timescales of minutes to hours, and it is a critical aspect of maintaining reliability. Balancing authorities, discussed in the Appendix, issue orders to generators to change their output as needed to maintain reliability. Balancing authorities can be utilities, or RTOs can act as balancing authorities in the regions where they exist. The rules for selecting which generators must increase or decrease output typically reflect an approach known as security-constrained economic dispatch (SCED). Under SCED, system operators ensure that electricity is produced at the lowest overall cost while respecting any transmission or operational constraints.[30] When generation from a low-cost source would jeopardize reliability, a higher-cost source is used. In other words, SCED has two goals: affordability and reliability.

SCED favors sources with low operating costs, and wind and solar sources do not have to pay for fuel. As a result, wind and solar sources typically generate the maximum amount of electricity they can at any moment.[31] Balancing typically involves quickly increasing or decreasing output from other sources in response to variable output from wind and solar sources. The capability to quickly change output is known as ramping, and electricity sources differ in their ramping capability.

System operators use a variety of electricity sources to balance generation from wind and solar sources. Some may be more commonly used in certain regions of the country, depending on local factors. Each has different benefits and limitations, some of which are summarized below.

- Reciprocating internal combustion engines (RICE) have seen an increase in installed capacity since 2000, partly in response to

[30] In EPACT05, Congress required FERC to study SCED and report its findings to Congress. That report is FERC, *Security Constrained Economic Dispatch: Definition, Practices, Issues and Recommendations*, July 31, 2006, http://www.ferc.gov/industries/electric/indus-act/joint-boards/final-cong-rpt.pdf.

[31] Exceptions do exist. Many such instances, known as curtailment, have been associated with limited transmission capacity.

higher levels of generation from wind and solar sources.[32] These sources have high ramping capabilities and use mature technologies. They usually use natural gas or fuel oil as fuel, so they have associated fuel costs and environmental impacts.[33]

- Steam turbines, usually fueled by coal or nuclear energy, have historically been operated at steady, high output levels, barring maintenance needs, because that is the most efficient and lowest cost operational mode for them. These sources are capable of ramping to some extent. This operational mode may provide revenue for certain sources located in regions of the country with low wholesale electricity prices. It might also result in higher costs for electricity from these sources, compared to when they are not ramped.[34]
- Wind and solar sources located in one area can balance wind and solar sources in other areas, since it is rare to have cloudy skies or calm winds over broad regions of the country simultaneously. This could have the benefit of using sources with zero fuel costs and zero emissions for balancing; however, existing electricity transmission system constraints limit the extent to which this is possible.[35]
- Energy storage can be used for balancing because it stores electricity during periods of high supply and then provides

---

[32] Suparna Ray, *Natural Gas-Fired Reciprocating Engines Are Being Deployed More to Balance Renewables*, Energy Information Administration, February 19, 2019, https://www.eia.gov/todayinenergy/detail.php?id=37972.

[33] Some other benefits and limitations are discussed in Chris Warren, "20 Power Companies Examine the Role of Reciprocating Internal Combustion Engines for the Grid," *EPRI Journal*, September 26, 2018, http://eprijournal.com/ start-your-engines/.

[34] Power plant operators do not often publicly discuss operational decisions because they can contain proprietary information. Aggregated and anonymous industry reports document how flexible operations like ramping are increasing. For example, see Peter Maloney, "How Market Forces Are Pushing Utilities to Operate Nuclear Plants More Flexibly," *Utility Dive*, October 4, 2016, https://www.utilitydive.com/news/how-market-forces-are-pushingutilities-to-operate-nuclear-plants-more-flex/427496/.

[35] Existing boundaries between balancing authorities can also pose hurdles to using remote wind and solar sources for balancing. Some Western states have created an Energy Imbalance Market to address such market hurdles. See Aaron Larson, "How Does the Western Energy Imbalance Market Work?," https://www.powermag.com/how-does-thewestern-energy-imbalance-market-work/.

electricity when supply is low. Many experts also see storage as a way to address the daily variability shown in *Figure 3* and thereby expand the utilization of installed wind and solar sources.[36] Many energy storage types are expensive and not currently deployed in large amounts.[37] Energy storage can be co-located with wind or solar generators, or it can be located at other sites in the power system or the distribution system.[38]

- Demand response, sometimes called demand-side management, involves adjusting electricity demand in response to available supply. This is counter to how the power system has historically been operated, but has become more commonly used.[39] Demand response includes programs in which electricity consumers voluntarily reduce their usage in exchange for financial compensation. Demand response can be a low-cost balancing option because it does not require electricity generation; however, it comes at a social cost because consumers do not use electricity at their preferred time.
- The electric power sector is working to improve the use of weather and power forecasting in system balancing. For example, MISO

---

[36] For example, "grid-scale storage helps with renewable integration, allowing higher renewable energy levels than would otherwise be possible." Ben Fowke, Chairman, President, and CEO of Xcel Energy in written testimony, U.S. Congress, Senate Committee on Energy and Natural Resources, *Hearing on Expanded Deployment of Grid-Scale Energy Storage*, 116th Cong., June 4, 2019.

[37] Current policy discussion is focused on how energy storage might balance daily variability of wind and solar sources, but some energy storage types could potentially balance seasonal variability as well. For a discussion of different energy storage technologies see CRS Report R42455, *Energy Storage for Power Grids and Electric Transportation: A Technology Assessment*, by Paul W. Parfomak.

[38] In Order No. 841, issued February 15, 2018, FERC defines an energy storage source as "a resource capable of receiving electric energy from the grid and storing it for later injection of electric energy back to the grid." The definition does not specify location, and FERC further clarifies that "electric storage resources located on the interstate transmission system, on a distribution system, or behind the meter fall under this definition." FERC, *Order No. 841: Electric Storage Participation in Markets Operated by Regional Transmission Organizations and Independent System Operators*, February 15, 2018, paragraph 29, https://www.ferc.gov/whats-new/comm-meet/2018/021518/E-1.pdf?csrt=5767993397659104309.

[39] Demand response in energy markets is discussed in CRS Report R43093, *Electricity Markets—Recent Issues in Market Structure and Energy Trading*, by Richard J. Campbell.

changed its wholesale electricity market rules in 2011 to create a Dispatchable Intermittent Resources program. This program allows wind sources to make use of their own generation forecasts and offer generation at five-minute intervals. Previously, offers had to be made on an hourly basis. This was creating inefficiencies in using wind sources since their output can vary over the course of an hour. Improved forecasting could result in increased use of low-cost wind and solar sources, but forecasting methodologies are still being optimized for this purpose.

The above considerations apply to bulk power system balancing today. Technological or policy developments could alter how system balancing is conducted in the future. Additionally, if wind and solar sources provided even larger shares of overall generation, new benefits or limitations for each balancing source type could emerge.

FEDERAL GOVERNMENT ACTIVITIES AFFECTING RELIABILITY AND BALANCING

Work at the federal level to address reliability needs associated with increased use of wind and solar sources has been underway for some time. For example, NERC created a task force in December 2007 to study the integration of VRE and identify gaps in reliability standards.[40]

The federal government undertakes actions in addition to the development and enforcement of reliability standards that affect electric reliability. FERC regulates interstate electricity transmission, which can be a key determinant of what sources are available to balance wind and solar. FERC also regulates wholesale electricity markets in most regions of the country. Market rules, including how SCED is implemented, can influence which individual generators are used for system balancing. Market prices

---

[40] NERC, *Accommodating High Levels of Variable Generation*, April 2009, https://www.nerc.com/files/IVGTF_Report_041609.pdf.

can directly affect project revenues and influence investment decisions. Additionally, Congress funds projects and programs that support technology development and deployment, including for sources and operations that improve reliability.

Some examples demonstrate the breadth of federal activities related to reliability.

- In EPACT05, Congress created Section 219 of the Federal Power Act that directs FERC to establish financial incentives for certain electricity transmission investments. FERC's resulting rule became effective in 2006 and includes provisions allowing higher rates of return, accelerated depreciation, and full cost recovery, all for investments and activities that FERC approves on a case-by-case basis.[41] Transmission investment has increased since the passage of EPACT05, although there may be many factors driving this investment.[42] On March 21, 2019, FERC opened an inquiry on potential changes to its transmission incentive policy.[43]
- In 2011, FERC issued a rule, Order No. 1000, revising requirements related to new transmission projects.[44] Among other revisions, Order No. 1000 increased the weight given to achieving public policy requirements when FERC considers approval of transmission projects.[45] An example of a public policy requirement might be a state requirement that a specified share of electricity sales come from renewable sources, a policy commonly known as

---

[41] FERC, *Order No. 679: Promoting Transmission Investment Through Pricing Reform*, July 20, 2006, http://elibrary.ferc.gov/idmws/common/opennat.asp?fileID=11089376.

[42] Lori Aniti, *Utilities Continue to Increase Spending on Transmission Infrastructure*, EIA, February 9, 2018, https://www.eia.gov/todayinenergy/detail.php?id=34892.

[43] FERC, "News Release: FERC Opens Inquiry on Improvements to Electric Transmission Incentives Policy," press release, March 21, 2019, https://elibrary.ferc.gov/idmws/common/OpenNat.asp?fileID=15192647.

[44] FERC, *Order No. 1000: Transmission Planning and Cost Allocation by Transmission Owning and Operating Public Utilities*, July 21, 2011, https://www.ferc.gov/whats-new/comm-meet/2011/072111/E-6.pdf.

[45] For a discussion see CRS Report R41193, *Electricity Transmission Cost Allocation*, by Richard J. Campbell and Adam Vann.

a renewable portfolio standard.[46] New transmission capacity is often needed to access and balance wind and solar sources.

- Several FERC orders demonstrate how market rules are changing in response to increased need for balancing and ramping. Order No. 745 allows demand response to earn compensation from wholesale electricity markets for providing energy services to balance the power system in day-ahead and real-time markets.[47] Order No. 841 allows energy storage systems to earn compensation from wholesale electricity markets for providing any energy, capacity, and essential reliability services they are capable of providing.[48] Implementation of Order No. 841 might lead to greater deployment of energy storage which could improve balancing.
- Various grant programs administered by the Department of Energy (DOE) have supported the development of new technologies that can balance wind and solar sources or support reliability in other ways. These include research and development into electricity generators; wind forecast models and methodology; power electronics for solar sources; and standards for interconnection into the bulk power system.[49]
- DOE's Office of Energy Efficiency and Renewable Energy (EERE) has funded research meant to improve short-term weather

---

[46] Twenty-nine states and the District of Columbia are implementing mandatory renewable portfolio standards, and an additional eight states have voluntary versions. Texas and Iowa have portfolio standards that set a requirement in terms of installed capacity, not share of electricity sales. DSIRE, *Renewable Portfolio Standard Policies*, October 2018, http://ncsolarcen-prod.s3.amazonaws.com/wp-content/uploads/2018/10/Renewable-Portfolio-Standards-2018.pdf.

[47] FERC, *Order No. 745: Demand Response Compensation in Organized Wholesale Energy Markets*, March 15, 2011, https://www.ferc.gov/EventCalendar/Files/20110315105757-RM10-17-000.pdf.

[48] FERC, Order No. 841: Electric Storage Participation in Markets Operated by Regional Transmission Organizations and Independent System Operators, February 15, 2018, https://ferc.gov/whats-new/comm-meet/2018/021518/E-1.pdf.

[49] The authorization and appropriations history for some of these programs is summarized in CRS Report R40913, *Renewable Energy and Energy Efficiency Incentives: A Summary of Federal Programs*, by Lynn J. Cunningham. Many of these efforts were organized within DOE as part of its Grid Modernization Initiative. Information on that effort is available at https://www.energy.gov/grid-modernization-initiative.

forecasting specifically related to wind power forecasts in two Wind Forecast Improvement Projects. DOE reports that advances made during this research include improved observations of meteorological data and improved methodologies for using those data in wind forecasts.[50]

POTENTIAL ISSUES FOR CONGRESS

Congress has held hearings related to the changes in the electricity generation profile of the country, and some Members raised concerns about reliability during these hearings.[51] Members may continue to examine reliability issues moving forward, in light of projections that wind and solar will become an increasingly larger share of electricity generation. For example, EIA's projection of existing law and regulations shows wind and solar sources contributing 23% of electricity generation in 2050.[52] Members may also choose to include reliability as part of any debate about policies to increase the generation from wind and solar sources.[53]

---

[50] DOE, *WFIP NOAA Final Report*, April 30, 2014, http://www.energy.gov/sites/prod/files/2014/05/f15/wf ipandnoaafinalreport.pdf and NOAA, *Wind Forecast Improvement Project 2 (WFIP 2) in Complex Flow*, https://www.esrl.noaa.gov/gsd/renewable/wfip2.html.

[51] See, for example, U.S. Congress, Senate Committee on Energy and Natural Resources, *Keeping the Lights On—Are We Doing Enough to Ensure the Reliability and Security of the U.S. Electric Grid?*, 113th Cong., April 10, 2014, S. Hrg. 113-271; U.S. Congress, House Committee on Energy and Commerce, Subcommittee on Energy and Power, *Hearing on Discussion Draft Addressing Energy Reliability and Security*, 114th Cong., May 19, 2015; U.S. Congress, House Committee on Energy and Commerce, Subcommittee on Energy, *Powering America: Defining Reliability in a Transforming Electricity Industry*, 115th Cong., September 14, 2017; and U.S. Congress, Senate Committee on Energy and Natural Resources, *Hearing to Examine the Electricity Sector in a Changing Climate*, 116th Cong., March 5, 2019.

[52] EIA, *Annual Energy Outlook 2019*, January 24, 2019, https://www.eia.gov/outlooks/aeo/pdf/aeo2019.pdf. Factors that could increase the use of wind and solar sources above this level might be missing from EIA's analysis. Such factors include new state or federal policies that might be enacted and unexpected cost declines for wind and solar sources or energy storage technologies.

[53] For example, H.Res. 109 in the 116th Congress states that accomplishing "Green New Deal goals" would require, in part, "meeting 100 percent of the power demand in the United States through clean, renewable, and zero-emission energy sources."

Preparing for higher levels of generation from wind and solar might require new approaches to maintaining electric reliability. The existing regulatory framework can accommodate some changes since FERC and NERC have authority to initiate development of new reliability standards. For example, NERC has raised the issue of whether it should develop new reliability metrics in light of the increasing use of VRE for electricity generation.

> In addition to its capacity supply assessment, NERC's Reliability Assessment Subcommittee should lead the electric industry in developing a common approach and identify metrics to assess energy adequacy. As identified in this assessment, the changing resource mix can alter the energy and availability characteristics of the generation fleet. Additional analysis is needed to determine energy sufficiency, particularly during off-peak periods and where energy-limited resources are most prominent.[54]

Congress could choose to provide guidance for FERC and NERC activities in this area.

Congress could also assess whether the existing regulatory framework is sufficient to maintain reliability if generation from wind and solar sources increase above current projections. One area of discussion is the siting and approval of transmission projects, particularly those that might result in enhanced availability of wind and solar sources for system balancing. Currently, the siting of electricity transmission facilities is largely left to the states. Section 1221 of EPACT05 directs FERC to issue permits for the construction or modification of transmission facilities in certain circumstances in areas designated by the Secretary of Energy as "National Interest Electric Transmission Corridors." This authority was to be exercised only if the relevant state agency lacks the authority to permit the transmission facilities or has "withheld approval for more than one year." Shortly after passage of EPACT05, DOE set out to designate the

---

[54] NERC, *2018 Long-Term Reliability Assessment*, December 20, 2018, p. 9, https://www.nerc.com/pa/RAPA/ra/Reliability%20Assessments%20DL/NERC_LTRA_2018_12202018.pdf.

National Interest Electric Transmission Corridors and FERC set up a framework for permitting transmission facilities on those corridors. However, federal courts vacated both agencies' actions, and neither agency has taken any significant action pursuant to their Section 1221 authority since that time.[55]

As noted above, most power outages occur on local electricity distribution systems, and these are regulated by state or local governments. Congress could consider expanding federal activities affecting distribution system reliability. This might involve studies of the factors (e.g., weather, aging infrastructure, VRE) that result in power outages. Such activities might also include federal financial support for projects or practices that improve reliability of distribution systems or encouraging new operational regimes such as independent distribution system operators (see earlier discussion of this issue in text box, "Distribution System Reliability").

Congress might also consider acting on the emerging and related issue of electric resilience.

Some support for an enhanced federal role in electricity system resilience exists. For example, the National Academy recommends

> Congress and the Department of Energy leadership should sustain and expand the substantive areas of research, development, and demonstration that are now being undertaken by the Department of Energy's Office of Electricity Delivery and Energy Reliability and Office of Energy Efficiency and Renewable Energy, with respect to grid modernization and systems integration, with the explicit intention of improving the resilience of the U.S. power grid.[56]

Many sources currently used to balance wind and solar have received federal financial support in the past, such as tax credits, grants to states or

[55] The finding against FERC was that it had inappropriately interpreted this authority as coming into force in cases where a state agency denied a permit. Piedmont Environmental Council v. FERC, 558 F. 3d 304 (4th Cir. 2009). The finding against DOE was that it had not sufficiently consulted with affected states prior to designating two National Interest Electric Transmission Corridors. California Wilderness Coalition v. U.S. Dep't of Energy, 631 F. 3d 1072 (9th Cir 2011).

[56] National Academies of Sciences, Engineering, and Medicine, *Enhancing the Resilience of the Nation's Electricity System*, 2017, p. 4, https://doi.org/10.17226/24836.

other entities, and DOE research programs. Congress might consider continuing or expanding this type of support if current activities affecting reliability were deemed insufficient.

APPENDIX. KEY RELIABILITY CONCEPTS FOR POLICYMAKERS

Electric reliability encompasses short-term and long-term aspects as shown in Figure A-1. System operators and reliability planners, governed by reliability standards from the North American Electric Reliability Corporation (NERC), have different practices in place to address reliability over these various timescales.

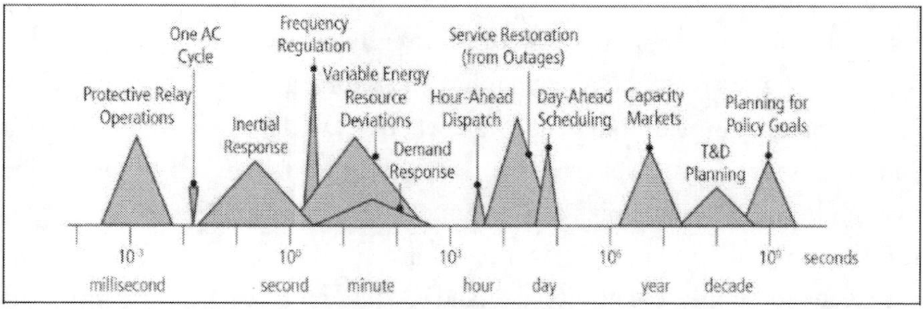

Source: DOE, Staff Report to the Secretary on Electricity Markets and Reliability, August 2017, p. 62, https://www.energy.gov/downloads/download-staff-report-secretary-electricity-markets-and-reliability.

Notes: AC = alternating current; T&D = transmission and distribution systems.

Figure A-1. Power System Reliability Timescales.

Reliability over Different Timescales

At the smallest timescales, typically seconds or less, are factors such as frequency control, voltage support, and ramping capability.[57] These are

[57] For more background, see FERC, *Reliability Primer*, 2016, https://ferc.gov/legal/staff-reports/2016/reliabilityprimer.pdf.

often automatic responses of power system components. NERC refers to these factors as Essential Reliability Services (ERS), and they are sometimes called ancillary services. Historically, many ERS were provided as a natural consequence of the physical operational characteristics of steam turbines. Wind and solar generators do not inherently provide ERS in the same way. They require additional electrical components to do so, and these are being more commonly deployed. In some cases, FERC has mandated the use of technologies that allow wind and solar to provide ERS.[58]

Balancing, described in the main body of this chapter, typically occurs over minutes to hours. Unlike ERS, balancing typically requires action by a system operator. Long-term aspects of reliability relate to planning for energy and transmission needs over months to years. This is sometimes referred to as resource adequacy. Policy goals, such as preferences for certain electricity sources over others, tend to influence long-term reliability planning more than shorter-term reliability aspects.

Planning for resource adequacy involves forecasts of electricity supply and demand. For variable renewable energy (VRE) like wind and solar sources, these forecasts require assumptions about wind and solar availability. Reliability planners commonly use planning reserve margins to assess whether planned generation and transmission capacity will be sufficient to supply electricity demand. A planning reserve margin is the difference between expected peak demand and available generating capacity at the peak period in each forecast year.[59] It is often expressed as a percentage where the difference is normalized by the peak demand value. According to NERC, reserve margins "in the range of 10-18 percent" are typically sufficient for ensuring reliability, although "by itself the expected Planning Reserve Margin cannot communicate how reliable a system is."[60] Reserve margins are calculated months or years ahead as part of

[58] For a discussion of the technical and market issues around ERS, see Department of Energy, *Staff Report to the Secretary on Electricity Markets and Reliability*, 2017, https://www.energy.gov/downloads/download-staff-reportsecretary-electricity-markets-and-reliability.

[59] NERC, *M-1 Reserve Margin*, accessed April 17, 2019, https://www.nerc.com/pa/RAPA/ri/Pages/PlanningReserveMargin.aspx.

[60] NERC, *Methods to Model and Calculate Capacity Contributions of Variable Generation for Resource Adequacy Planning*, March 2011, p. 7, https://www.nerc.com/files/ivgtf1-2.pdf.

assessments of whether and where reliability concerns might exist. High planning reserve margins may indicate a likelihood that reliability will be maintained, but, especially when variable sources are present, they might not be predictive. That is, a high planning reserve margin does not guarantee reliability and a low planning reserve margin does not guarantee power disruptions. At the national level, NERC annually assesses resource adequacy over a 10-year forecasting window. NERC uses historic VRE generation data in its assessment and has noted "methods for determining the on-peak availability of wind and solar are improving with growing performance data."[61] In its 2018 Long-Term Reliability Assessment, NERC recommended enhancing its reliability assessment process to account for events, like those noted in the "Changing Electricity Generation Profile" section above, during which VRE sources provided large shares of generation during off-peak periods.[62]

Solar eclipses, though rare events, provide opportunities to test the ability of grid operators to reliably operate the grid when solar sources are unavailable. The August 21, 2017, solar eclipse that affected many parts of the United States was one such opportunity. According to NERC, no reliability issues developed during the event, in part because of the measures implemented in advance by the electric industry.[63]

Electric Reliability Regulatory Framework

Current electric reliability planning is a coordinated process involving multiple entities and spanning multiple jurisdictions. These reliability planning organizations share responsibility for, among other responsibilities, ensuring electricity from wind and solar sources are

---

[61] NERC, *2017 Long-Term Reliability Assessment*, March 1, 2018, p. 5, https://www.nerc.com/pa/RAPA/ra/Reliability%20Assessments%20DL/NERC_LTRA_12132017_Final.pdf.

[62] NERC, *2018 Long-Term Reliability Assessment*, December 20, 2018, p. 9, https://www.nerc.com/pa/RAPA/ra/Reliability%20Assessments%20DL/NERC_LTRA_2018_12202018.pdf.

[63] NERC, *State of Reliability 2018*, June 2018, p. 38, https://www.nerc.com/pa/RAPA/PA/Performance%20Analysis%20DL/NERC_2018_SOR_06202018_Final.pdf.

reliably integrated into the power system. *Table A-1* summarizes these entities and their responsibilities.

Table A-1. Selected entities involved in power system reliability

Entity Name	Selected Reliability Activities	Enforcement Mechanism	Oversight Mechanism	Notes
Federal Energy Regulatory Commission (FERC)	Approves new or modified reliability standards proposed by NERC and conducts enforcement activities.	FERC can enforce compliance plans, reliability enhancements, and impose civil penalties for violations of reliability standards.	Commissioners are nominated by the President and confirmed by the Senate.	Pursuant to EPACT05, FERC may approve reliability standards that are developed by NERC. FERC rules for investigating potential violations of reliability standards are codified in 18 C.F.R. §1b
North American Electric Reliability Corporation (NERC)	Sets reliability standards for industry and annually assesses long-term (10- year) resource adequacy.	NERC files Notices of Penalty with FERC.	Independent board oversees NERC. FERC may direct NERC to revise reliability standards.	FERC certified NERC as the Electric Reliability Organization (ERO) in 2006. EPACT05 specifies duties of the ERO regarding reliability standards
Regional Entity	Monitors and enforces compliance with NERC reliability standards.	Can recommend penalties on entities that violate reliability standards and/or fail to mitigate any violations.	NERC can hear appeals in case of disputed violations.	EPACT05 allows NERC to delegate some responsibilities to Regional Entities. Older documents may refer to eight regional entities, but in 2018 the Southwest Power Pool Regional Entity ceased operations and its members transferred into other regional entities, leaving seven regional entities.

Entity Name	Selected Reliability Activities	Enforcement Mechanism	Oversight Mechanism	Notes
Balancing Authority (BAs)	Maintains real time power system reliability in response to system conditions.	Enforcement conducted by appropriate Regional Entity.	Mostly conducted at the local or regional level.	BAs as separate entities are not required by law. Instead, NERC reliability standards require BAs to ensure that electricity supply and demand are balanced in real time. Entities voluntarily register to take on these duties. All of the RTOs act as BAs.
Public Utilities Commissions (PUCs)	In some states PUCs approve power system investments, including those related to power system reliability. PUCs have exclusive jurisdiction over distribution system reliability.	Could include imposing penalties or denying cost recovery for damages associated with poor reliability.	Some state utilities commissioners are appointed by governors; others are elected.	Some state policies, including those supporting wind and solar energy development, might have consequences for reliability. Some states are considering adoption of performance-based rates to provide financial incentives to utilities with positive reliability performance.

Source: CRS analysis.

Notes: Other entities, including generator operators, transmission operators, and reliability coordinators are involved in reliability planning, but their activities are not shown here. Each of those entities has a different scope of function in maintaining power system reliability. A full list of the roles of power system entities in maintaining reliability is available in the NERC Reliability Functional Model at https://www.nerc.com/pa/Stand/pages/ functionalmodel.aspx.

In the Energy Policy Act of 2005 (EPACT05; P.L. 109-58), Congress gave FERC responsibility for reliability of the grid through the setting and enforcement of electric reliability standards.[64] These standards are developed by NERC and approved by FERC in the United States.[65] NERC

---

[64] 16 U.S.C. §824o.

[65] 65 Due to the integrated nature of the bulk power system, NERC oversees electric reliability for parts of Canada and Mexico as well. The electric industry originally established NERC's predecessor, the National Electric Reliability Council, in 1968 to coordinate voluntarily reliability efforts. Alaska and Hawaii are exempt from NERC's jurisdiction.

has set over 100 reliability standards that cover all timescales of reliability planning.[66]

Congress gave NERC authority to enforce reliability standards in EPACT05. Per statute, NERC has delegated this authority to the Regional Entities shown in Figure A-2. The jurisdiction for enforcing compliance with reliability standards includes "all users, owners and operators of the bulk-power system" within the contiguous United States.[67]

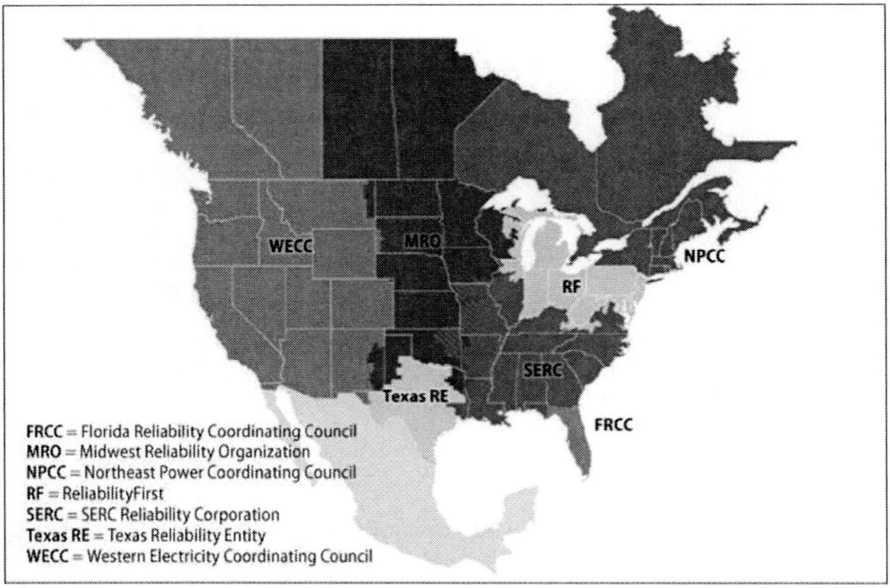

Source: CRS, adapted from NERC, "Key Players," accessed April 17, 2019, https://www.nerc.com/AboutNERC/ keyplayers/pages/default.aspx.

Notes: The multicolored areas represent areas where some entities are members of MRO and some entities are members of SERC. Alaska and Hawaii are not shown because mandatory reliability standards do not apply in those states.

Figure A-2. NERC Regional Entities.

---

[66] The number of standards in effect changes as NERC modifies and creates standards in response to directives from Congress or FERC and changing conditions in the electricity system. The current list of standards is available at https://www.nerc.com/pa/Stand/Pages/ AllReliabilityStandards.aspx?jurisdiction=United%20States.

[67] 16 U.S.C. §824o(b)(1).

Separate from the tasks of setting and enforcing reliability standards is the task of reliably operating the power system in real time. Per NERC's reliability standards, balancing authorities carry most of the responsibility for matching generation levels with electricity demand. Balancing authorities can have different geographic footprints. RTOs act as balancing authorities and they may have a footprint spanning multiple states. Other balancing authorities might have a footprint spanning an area within a single state. Another class of entities with operational responsibilities are reliability coordinators. A reliability coordinator may operate over larger geographic areas than balancing authorities and can overrule real-time decisions by balancing authorities to preserve the larger scale power system reliability. RTOs typically also act as reliability coordinators. NERC has certified 66 balancing authorities and 11 reliability coordinators in the United States.[68]

---

[68] NERC, *NERC Balancing Authorities*, updated October 1, 2015, https://www.nerc.com/comm/OC/RS%20Landing%20Page%20DL/Related%20Files/BA_Bubble_Map_20160427.pdf. In this map, balancing authorities are denoted by abbreviations, and reliability coordinators are denoted by colored areas. The multicolored areas on the map represent regions where balancing authorities receive reliability coordinator services from either of two reliability coordinators.

INDEX

C

D

E

F

G

H

I

M

N

Q

R

S

V

W

Y

Z